Logistics

Logistics

The Key to Victory

Jeremy Black

Pen & Sword
MILITARY

First published in Great Britain in 2021 by
Pen & Sword Military
An imprint of
Pen & Sword Books Ltd
Yorkshire – Philadelphia

Copyright © Jeremy Black 2021

ISBN 978 1 39900 601 9

Typeset by Mac Style
Printed and bound in the UK by CPI Group (UK) Ltd, Croydon, CRO 4YY

Pen & Sword Books Limited incorporates the imprints of Atlas, Archaeology, Aviation, Discovery, Family History, Fiction, History, Maritime, Military, Military Classics, Politics, Select, Transport, True Crime, Air World, Frontline Publishing, Leo Cooper, Remember When, Seaforth Publishing, The Praetorian Press, Wharncliffe Local History, Wharncliffe Transport, Wharncliffe True Crime and White Owl.

For a complete list of Pen & Sword titles please contact

PEN & SWORD BOOKS LIMITED
47 Church Street, Barnsley, South Yorkshire, S70 2AS, England
E-mail: enquiries@pen-and-sword.co.uk
Website: www.pen-and-sword.co.uk

Or

PEN AND SWORD BOOKS
1950 Lawrence Rd, Havertown, PA 19083, USA
E-mail: Uspen-and-sword@casematepublishers.com
Website: www.penandswordbooks.com

For
Simon Heffer

Contents

Preface

The military cannot operate without supplies as every commander knows. Yet, literature on the subject, while valuable, is far less common than on many other aspects of war, notably weaponry, tactics and command. This book sets out to address a gap. To offer a complete account would take a lifetime, but my plan is different. I focus on key conflicts, developments and concepts in order to show that, although the technology has changed, as it will continue to do, the underlying issues remain the same.

It is appropriate at the start to turn to the major text in the field, Martin van Creveld's *Supplying War,* and to ask how the situation might look differently compared to when it was finished in 1976 and first published in 1977. The book itself was a major success, praised in a range of publications, recommended to officers in a number of states, notably the United States, and frequently reprinted, including in 1980, 1982, 1983, 1984 (twice), 1985 (twice), 1986 (twice), 1987, 1990 (twice), 1991, 1992, 1993, 1994 and 1995. In 2020, it continues to sell well. Clearly written and based on a good range of material, the book is still worth reading, but it also, and understandably so, very much looks of the moment when it was produced.

That it does so throws much light on the changing nature of logistics itself. First, and most seriously, there is the geographical context, that in van Creveld, China, India and Japan are forgotten, as is Latin America. Africa is only there for the Allies and Axis to fight in North Africa in 1940–2, essentially in order to discuss the Afrika Korps. Antiquity and the Middle Ages of course do not exist, and the book begins with 1560–1660, the period of Michael Roberts' supposed Military Revolution. Navies and air forces are not a matter of interest.

To put it mildly, all of these exclusions are highly problematic. But so is the focus in what is covered. The focus is fundamentally on Napoleon, Prussia and the two world wars, and with the latter handled in a somewhat strange fashion in that, in the discussion of the Eastern Front in 1941–5, the Soviet Union plays scant role as an active power. Instead, it is German inadequacies that are to the fore. This, moreover, is not the sole problem with what is covered. Van Creveld's attention is on the big battalions, with the criteria for their maintenance on the offensive being his concern. That provides no opportunity to consider the full spectrum of military operations, even only on land, even only by Western powers, even only in warfare between them.

The challenge therefore is clear – to offer an account of logistics that is broader in its chronological, geographical and thematic ranges, engaging with the full spectrum of conflict, and seeking to avoid a teleological approach, and at the very least to offer a study that is fit for purpose, the key element in logistical capability and practice. I offer a broad conceptualisation of logistics and approach change through time without falling into the trap of progress. Instead, logistics is considered in relation to military cultures and as a set of dynamic responses to specific conditions. There is a standard confusion of logistics with technology, and an assumption that technology can wipe away all problems. Instead, as this book will make clear, technological and organisational developments, while they solve some problems, also create new ones.

I am most grateful to Leopold Auer, Stephen Badsey, Ahron Bregman, Enrico Cernuschi Ong Wei Chong, Mike Dobson, John France, Heiko Henning, Caleb Karges, Luigi Loreto, John Lynn, Tim May, Stephen Morillo, Praytyah Nath, Steven Pfaff, Kaushik Roy, Mike Sibley, Mark Stevens, and Ulf Sundberg for commenting on all or part of earlier drafts. I have benefited from advice on specific points by Rodney Atwood, Peter Caddick-Adams, Mike Broers, Mike Cailes, William Callaway, Ted Cook, Kelly DeVries, Mike Duffy, Mark Fissel, Dick Goodenough, Bob Higham, John Keenan, Hilton King, Roger Knight, Nick Lipscombe, Graham Lord, Scott Myerly, Gervase Phillips, Paul Rahe, Frédéric Saffroy, Nigel Saul, Mark Stevens, Claudio Vacanti, Sharkey Ward and Everett Wheeler. None is responsible for any errors that remain. Lester Crook has proved a really helpful editor. It is a great pleasure to dedicate this study to Simon Heffer, a fellow labourer in the groves of Clio and a man of judgement, scholarship, sage advice and good cheer.

Abbreviations

Add.	Additional Manuscripts
AE	Paris, Archives du Ministère des Affaires Etrangéres
AWM	Canberra, Australian War Memorial, Archives
BL	London, British Library
CAB	Cabinet Office papers
CP	Correspondance Politique
CRO	County Record Office
HL	Huntington Library, San Marino, California
IO	India Office papers
JMH	*Journal of Military History*
LC	Washington, Library of Congress, Manuscript Division
LH	London, King's College, Liddell Hart Library
LOC	Logistics Operations Center
NA	London, National Archives
NAM	London, National Army Museum
NMM	London, National Maritime Museum
RFA	Royal Fleet Auxiliary
SP	State Papers
STUFT	Ships Taken Up From Trade
WO	War Office papers

Unless otherwise stated, all books are published in London.

Introduction

'This definition of logistics encompasses the broad range of sustainment activity, from production to consumption, movement, deployment and withdrawal.'

Army Doctrine Publication. Logistics (1996).[1]

'A scrimmage in a Border Station -
A canter down some dark defile
Two thousand pounds of education
Drops to a ten-rupee *jezail* [Afghan musket].
The Crammer's boast, the Squadron's pride,
Shot like a rabbit in a ride!

....

With home-bred hordes the hillsides teem.
The troopships bring us one by one,
At vast expense of time and steam,
To slay Afridis where they run.
The "captives of our bow and spear"
Are cheap, alas! as we are dear.'

From Rudyard Kipling 'Arithmetic on
the Frontier' (1886), referring to the British
on the 'North-West Frontier' of India.

A Cinderella subject, logistics is more usually cited than discussed at length, and there is only limited theoretical discussion of logistics, compared in particular with that on strategy. Moreover, what is covered in the subject varies greatly. The obvious component, and that popularised by Antoine-Henri Jomini, the leading commentator of Napoleon, when he used the term in chapter six of his *Précis de l'Art de la Guerre* (1838), is the material aspects of logistics: getting men, food, munitions and other *matériel* to the right place at the right time, and, more particularly, the interface between the practices and institutions that sustain military endeavours and the geographical challenges with which they contend.[2] Jomini took it from *logis* or 'lodging place' in French usage. There was also, derived from Greek, the French mathematical term *logistique*. The word 'logistics' was used in English from 1846. Jomini, however, focused on the art of moving armies rather than supplying them, and it was not until the 1930s that the word came back into French military vocabulary, and then from the Italian 'logistica' popularised during the Italian-Ethiopian War of 1935–6. The British Army term for logistics was 'supply', which was under the Quartermaster General (a post that no longer exists). The term by the mid-twentieth century was 'Administration' or 'Admin,' which embraced supply and maintenance. In turn, the Royal Logistic Corps was founded in 1993.

The Swiss-born Jomini (1779–1869) had an unusual background as a general, having been to a business school and worked, from 1795, as a banker before becoming secretary to the Swiss Minister of War in 1798, acquiring experience in organisation. He was a staff officer under Marshal Ney in 1805, and thereafter with Napoleon himself in 1806–7. Transferring to Russian service, he became a major military commentator, and logistics was to be fully covered in his works. They were highly influential, not least in the United States where the first English translation appeared in 1854.[3]

Opinion as to how far back in the supply chain attention should go varies greatly in the literature. So also with the provision and movement of other services, notably medical services and sex. Were, for example, the women forced into prostitution for troops not an imprisoned part of the logistical chain? There is also in evaluating logistics its importance to sustaining morale, and hence fighting capability, and, therefore, the question of how much weight to place on that factor. Moreover, the cultural factors not only behind success or failure in logistics, but also in defining logistics and this success or failure, are of great significance, but, again, it is unclear how best to evaluate this factor. What, however, would be a mistake would be to neglect those elements that could not be quantified.

Logistics, to generalise, can be discussed in terms, first, of the art of raising and, especially, maintaining armed forces. Secondly, comes logistics and the natural world, both in terms of resource distribution, and with reference to environmental pressures on supply systems. Thirdly, comes logistics and underlying economic systems, in terms of agriculture, industry, and extant network capabilities, whether merchant, capitalist, cultural or in other forms, including levels of monetarisation. Fourthly, comes logistics and government processes. Fifthly, logistics and the nature and scope of military tasking are at issue. Then it is necessary to add variations at strategic, operational and tactical levels, and with reference to regular and irregular warfare, land, sea and air capabilities, and the particular circumstances of individual physical, military and cultural environments.

The intention here is to offer a conceptual introduction that is designed to begin a debate. The thesis is that logistics should be approached in terms of the idea that, rather necessarily than improvement through time, you have adaptation with reference to fitness for purpose, the fitness provided by task, environment and means, the last involving political culture in the shape of governmental type and related military systems, as much as sources of supply. The recent and continuing war in Afghanistan provides an example of logistical means being dictated for all sides by the assumptions of political culture as much as supply considerations. Moreover, the costs of protecting logistical systems drives up the logistical burden to the supposedly 'better' power.

This contextual approach is a matter therefore of specificities, and not some linear development; and administrative organisations and practices take on their value in terms of the ability to respond to these specificities, and not in accordance with a model assessed with reference to an asserted universal or quasi-universal value and applicability. The emphasis on specificity includes that of the frequently misleading context of subsequent scholarly assessment, an assessment often delivered in terms of judgment, as in x was backward and y impressive, with the scale somehow again of universal validity in both time and space.

Indeed, in addition to the issues involved in the response to logistical practice to past circumstances, comes those of the response in the assessment of logistics to subsequent political assumptions. For example, while valuable, Géza Perjés' oft-cited account of the provisioning of European armies in the second half of the seventeenth century, was in fact very much linked to a Marxist analysis. In part, this analysis was related to the idea of the imposition of a 'second serfdom', although it was also linked to world systems approaches critical of capitalism:

> the numerical growth of armies far exceeded progress in the means of production... This advance of politics beyond military factors was a natural consequence of the world-wide struggle for hegemony in Europe, for the monopolisation of world trade and the colonies; but that such advance could take place at all was made possible by a growing absolutist centralisation...[4]

In practice, this account, like that by Michael Roberts on the supposed early-modern European military revolution, was very much of its time and place, while Perjés' subsequent argument that strategy had reached a state of crisis due to logistical problems,[5] while convenient to repeat, also underplayed the military achievements of the period as well as adopting an overly-simplified account of war. For an instance of the war, the indecisiveness of the War of the Spanish Succession in the Low Countries was mentioned by Perjés, but not its decisiveness in both Italy and Spain, leading respectively to Habsburg and Bourbon dominance. And so on.

The key instance of assessment in terms of subsequent political assumptions that is still to the fore, an assumption, however, that now is increasingly challenged,[6] is the argument that provision by the state was inherently more rational, and easier to plan and to coordinate with the armed forces, and, therefore, in some way, at least functionally, better. This was very much the argument not only seen with war as national mobilisation, and also with authoritarian and totalitarian states, but also with theses of modernisation in which the state plays a key enabling role, and notably those of the relationship between war and the state,[7] with mobilisation for war as a cause of modernisation. Whether logistics is defined in terms of the transportation of supplies, or with reference to a broader state capacity perspective, these conceptual and methodological questions are to the fore.

The ambivalence of modernity as a concept, however, is readily apparent. An example is provided by the issue of conscription, one that had serious implications for logistical requirements and, more generally, both reflected and contributed to an ethos of national mobilisation. Yet, the building blocks of the analysis are more problematic than sometimes argued. For example, the conscription of Revolutionary and Napoleonic

France following the *levée en masse* introduced in 1793 is regarded by some commentators as harnessing nationalism and leading towards modernity, representing a revolution in both military affairs and the state;[8] on the other hand, conscription, of people and goods, including draught animals, was scarcely a monopoly of modern and modernising states, or at least of the conventional listing of these states, both at the time (for example Burma), and across history. Moreover, on a pattern seen before, for example, both in the later Roman Empire and in the later Middle Ages, Western states after the final defeat of Napoleon at Waterloo in 1815 preferred for several decades to rely on professional long-service regulars, rather than large numbers of conscripts; only for the latter method to come to the fore in the second half of the nineteenth century. In addition, throughout, the professional long-service regulars proved the pattern for the British army, the most successful multi-purpose force of the period. Indeed, features other than conscription can be regarded as equally, if not more, 'modern', namely those of capitalist professionalism; a contrast in recruitment that can be seen across history.

If there is such a lack of certainty in this respect as to what constitutes modernity, it is unclear why the same should not be the case of other aspects of military activity, including logistics. So also in terms of a consideration of the situation with reference to two axes: first, that of the exercise of formal state power and, secondly, the role of choices in a free economic market.[9] These axes have clear consequences for both recruitment and logistics, but it is difficult to see how modernity can be readily located in this context. The increase in contractor services in the American military has been particularly apparent from the 1990s.[10] Earlier, most of the infrastructure used by the Americans during the Vietnam War was built by RMK-BRJ, a consortium of American companies. This process can be regarded as anti-modern, in that the state becomes dependent on private concerns, but, alternatively, that can be regarded as a form of specialisation of provision within a state-directed project; and thus as an aspect of the development by and towards specialisation praised by Adam Smith in *The Wealth of Nations* (1776).

Thus, returning to the late eighteenth and early nineteenth centuries, the combination, in India, of an already-established large-scale military labour market, able to respond to new challenges and opportunities in tasks and training, with, in the case of Britain, the finances of a dynamic sector of the world economy, produced a volatile situation in which new outcomes were readily generated as part of the problems and solutions of hybrid systems. The British benefited, their campaigns drawing on the resources of the East India Company, the British government, British 'Agency Houses' (local trading and finance establishments), such as John Palmer and Co., and Alexander and Co., and, most significantly, Asian financial networks and supplies. In the end, the military fiscalism of British India was to triumph,[11] although local powers posed a considerable challenge to the British into the early nineteenth century.[12]

Furthermore, as is now argued, not least by British historians, if military-fiscal systems, or military fiscalism, are seen as the key element toward both effective military support and state formation,[13] this development was both far from novel and, moreover, rested on broader social, economic and political arrangements.[14] Linked to this, the relationship between social foundations and mobilisation varied greatly. Thus, in Burma under 'Alaungpaya (r. 1752–60), a dynamic expansionist, the permanent professional military (as opposed to wartime conscripts) was supported by the provision of state land, and soldiers were obliged to grow their own food. Burma benefited from the intense paddy cultivation in the Irrawaddy valley, which generated a huge amount of rice that both provided Burma with a large population and enabled the Burmese monarchy to maintain a large number of troops.

This issue returns us anew to another point about logistics which is the extent of the supply system that is in discussion. Is it a case simply of the provision of supplies, generally food, water, fuel, and munitions, to the military in the field, with, in addition for modern armies, information systems, medical support and maintenance facilities, especially for the heavy vehicles, aircraft and helicopters on which Western armies currently rely so heavily? If so, how pertinent is it to keep 'force generation', the crucial supply of manpower separate, as recruitment does not exhaust that subject? Indeed, in light of the often very heavy depletion due to casualties and disease, battle casualty replacements are an important aspect of logistics, as they have been throughout history, not only in times long past but also more recently. So also with medical support as a means of returning troops to combat availability, a key element for example in the First World War, and also as an aspect of preserving morale, as for example with the significance of casualty care, withdrawal and subsequent treatment for Western forces in Afghanistan and Iraq in the early twenty-first century.

More generally, supply in the field is but part of a supply system and, therefore, logistics is as much to do with food, fuel and weaponry provision as part of the general preparedness of the military, as it is to do with their campaigning availability, whether at the operational or the tactical level.[15] In theory, there is a distinction, like, for example, that between strategy and policy, but, in practice, this availability depends on the preparedness and cannot readily be detached from it. Separately, as also with strategy and geopolitics, there are the conceptual and methodological issues of applying the idea of logistics as distinctive to much of history when that distinction did not pertain.

Furthermore, this point interacts with the differing levels of war, notably tactical, operational and strategic. This interaction includes the general point about resource availability being a central element of logistics capability, and thereby directs the subject from the operational support-for-a-campaign level that appears to be all too common as the sole focus of logistics when the subject is discussed, to that of the past understanding and usage of capability. Linked to this point, the term 'supply lines' carries different meanings depending on the environmental, technological and economic contexts. There were also lines of supply at different levels of warfare: strategic, operational and tactical. At the strategic level, these lines, to an extent, were coterminous with systems of production and trade, with the important caveat that these were different in wartime, not only for specific military reasons, but also due to the politics of war, in the shape of alliances, and the economics of war. The operational level is the model of need and response in most people's heads when they hear the term. The tactical level covers topics such as access to water for armies; for navies, such access was more operational an issue due to the difficulty of access to drinking water.

In practice, the tactical level of logistics is often the most significant as far as conflict is concerned, but also the one that receives least attention. In part, this is because the source level is weakest. In part, this was because, across history, there was frequently a poor level of institutional provision of supplies to the military in the field and, instead, a focus on their part on *ad hoc* responses to need. This was a form of subsistence logistics that, to a degree, was a parallel with the subsistence agriculture of the period. The consequences are that, for most of history, much of the practice of logistics at the tactical level focuses on activities aimed at the acquisition of supplies that were immediate or short-term, and also highly specific to the spot. The most significant activities were obtaining food and forage at the roadside, and water at the spring or river, a practice that could be extended to include money, sex, and other goods and services that were seized, obtained, or purchased; the processes involved frequently overlapped and left scant evidence in the literature. An exception, sometimes, was the use of military justice against those whose behaviour was thought unacceptable. A particular consequence that often leaves indirect evidence in the form of the number of the sick was drinking water from water courses that rapidly became polluted or that were already problematic.

Need was not only a matter of supplying units on the move but also, very differently, stationary ones, including, in particular, garrisons.[16] In the latter case, there was not the prospect of depredations in new areas in order to obtain supplies, but, instead, often the need to provide a system of some regularity, albeit one that could vary with summer and winter quarters. Such a system was an element of the predictability that aided units, not least by diminishing the need for the search for supplies. However, care is required before arguing therefore that a more systematic provision of supplies was required in order to 'free' troops from this search for supplies. That thesis both downplays the operational, strategic and political value of such a search, in weakening opponents (both foreign and domestic) and corroding their willingness to fight, and also implies that such a systematic provision was both possible and cost-efficient in terms of the resources available. That might have been the case, but was not necessarily so.

Sieges of garrison positions were themselves often a critical instance of the confluence of levels of war and logistics. Many forts were there as part of logistical systems, in terms of being a protected gathering point of regularly-obtained supplies, protected by the fortifications and by the garrison, thus creating its own logistical challenges. If, because the gathering point gained significance, an attacker decided to take it and had to rely on a siege, then the attacking force had to give up its mobility and create, on the spot, some system of supply in order to be able to sustain operations.

There was also the element of seizure and/or purchase as aspects of a free, or rather freer, market provision that captures the role of raw enterprise in the system. That was more the case at the tactical than the operational level, but, nevertheless, was found at the latter and to devastating effect.[17] Alongside coerced contributions, pillage could be very important to the morale of troops and accepted as such, as in the Ottoman army which seized women as sex-slaves in the fifteenth century, as the Japanese military was to do in the Second World War. This, however, created a problem when the Ottomans moved onto the defensive, contributing, alongside military failure, to mutinies in 1717–19 and 1730. The role of pillage, including sexual pillage in the form of rape, can also repeatedly be seen in operations by irregular forces in recent decades, for example by the Janjaweed militia in Darfur, by Boko Haram in Nigeria, and by ISIS with Yasidi women in Iraq. Such activity in part is a way of achieving psychological mastery over opponents.

In terms of the emphasis on cultural spaces, or strategic cultures, there is also the need to understand that functional best-practice accounts of logistics have to yield to the specificities of particular understandings of appropriate and necessary conduct, these understandings being an aspect of what is known as strategic culture in the sense of longstanding assumptions of individual states.[18] This point about particular understandings covers the entire ground from looting to sustenance levels, and thus undermines abstractions such as sixteenth-century logistics or, even, French sixteenth-century logistics, because what was the desired and/or acceptable practice in civil warfare, and notably in suppressing heterodox rebellion, as during the French Wars of Religion (1562–98), was different to that for France in fighting abroad. Techniques might be the same, but not generally practice.

Once the element of specific strategic culture is pushed to the fore, then the practice of adopting or adapting economic concepts such as opportunity costs, cost-benefit analysis, and marginal returns, appears of limited value,[19] while the narrative of technological change, and notably of its dominant role, as the key enabler of logistical improvement is at least qualified. It is of course useful to see how parameters changed with steam power or the internal combustion engine or aircraft, but that approach also has drawbacks. These can include underplaying the weaknesses and (separately) costs of these new systems, and the ability of opponents to undermine and/or match them as well as the need to adapt to existing systems to help make new ones work.

A focus on technological capability can also provide a horizontal (or over-the-horizon look) that says little about the reasons for choosing or accepting particular logistical outcomes, both with reference to the use of technological possibilities and in terms of the politics of control sought. With the politics of control, it is not solely a matter of the relationship with civilians, both in the sphere of combat and in the metropole (including the regulation of sexual behaviour),[20] but also the nature of state, indeed military, control over armed forces, an element that has varied greatly. Logistical capability therefore can, at least in part, be related to the terms, indeed conditionality, of service and support.

Here, indeed, there is the issue of present-day variety, for the logistics of today are as much a matter of warlords in the Central African Republic or Lebanon supporting their militias, as of American troops across the world eating standard rations or MREs (meals ready to eat). Whatever our preferences and norms, and academics are prone to favour bureaucratic efficiency,[21] one form is not more fundamental, functional nor progressive than another; each represents a response to tasks, ethos and circumstances. Moreover, access to weapons, money and supplies (notably food), is an aspect of power whatever the logistical system, and so also with the patronage that is deployed and the clientage system that is called on through the provision of military support.[22] The access and patronage frequently interact to determine not simply the operation of the logistical system, but also the very system itself. The related multiple character, indeed confusion, of circumstances in the present can be readily extrapolated to include the past, and it is disappointing that that has not become more common as a replacement for often unitary models of historical development.

The contrasting international responses to the 2020–1 Covid-19 pandemic has thrown to the fore the question of the respective value of dictatorial and democratic societies, and has done so in a fashion that is widely critical of the free nature of globalising exchanges. That approach, however, does not necessarily mean that authoritarian systems (and also militarised or, at least, military-led ones) are more appropriate at present. Nor were they necessarily so in the past, when there were also differences between more and less controlling practices, with mercantilism, for example, contrasted with free market capitalism. Consideration of present and past circumstances is not co-determinant, but it would be naïve to believe that they are not linked in assessment by scholars and others. Thus, views on logistics in part can take us to the debate about the optimal type of government, one certainly going back to the Greeks.

Linked to this comes the point made by Edmund Burke, an opposition parliamentarian in the British House of Commons in 1772. Having informed the House, in the misleading Eurocentric fashion of the day, that:

> The practice of keeping on foot large standing armies in time of peace, though not absolutely modern, (for we read of such an institution in ancient times), is new to the extent it is now carried in Europe. Charles the 5th [Holy Roman Emperor, r. 1519–56] was perhaps the first great monarch that set the example.

Burke added that such a course exhausted the state.[23] Thus, as Napoleon was unwittingly to demonstrate, logistical success in the narrow sense, while operationally effective, was only a short-lived panacea. Indeed, such success could be presented as potentially destructive, and, thereby, a strategic failure. Again, that point was contingent to a particular political context, but also worthy of wider consideration.

These points do not exhaust the range of preliminary theoretical reflections. Others include the contrast between land and naval logistics, (again understood as a response to specific requirements), and the discussion of scale. Moreover, the idea of development through time needs to address the question of whether Han China or Classical Rome should be considered less impressive than later[24] or modern counterparts; the same for 'non-state actors' of the period. And on sea as well as land, so with the complex logistics involved in the mobilisation of Mediterranean galley fleets from Antiquity to the seventeenth century,[25] as with the crucial role for the logistics of the Rome fleet in the First Punic War (264–241 BCE) of Syracuse and Sicilian other allies. The significance of maritime links for the movement of goods and people, both military and otherwise, was such that control over and from them offered a strategically and logistically cohesive core, as with the Ottoman focus on the 'Black Sea, Marmara and Aegean transportation, communications and market exchange nexus'.[26] Repeatedly, the need to integrate land and sea logistics ensures that division in their treatment is very unhelpful.

Much general discussion of logistics (as opposed to valuable specific research) leaves out or underplays the Ancient World and, often, non-Western societies, so that privateering for example, a classic form of contracting for naval support, is covered largely only if by Western powers.[27] Due to this situation, perceptions about modernity, periodisation, the value of societies that keep records, notably statistical ones, and technological triumphalism, all play a major role in the understanding of logistics.

Much of the standard work on logistics indeed deals essentially with the last half-millennium with the classic work, that by Martin van Creveld, cited by publishers as a reason for not needing another general work on the topic.[28] This was very much an account that drew on standard chronologies and analyses of a move toward total and industrial warfare, chronologies and analyses that suffered from a failure to devote due attention to conflict prior to 1792, to non-Western warfare, and, as it turned out, that after the Cold War. However, some valuable collections on logistics looked more widely,[29] and there is a wealth of excellent recent material on earlier periods. Indeed, the range and quality of such work encourages a rethinking of the standard approach.

There are significant studies of logistics in the Ancient world. Moreover, recent years have seen the appearance of much on medieval Europe, including, taking forward the use of *Sachkritik* (source criticism) by scholars such as Bernard Bachrach, the use of logistical modelling techniques that help by means of equations to evaluate total needs and the related capability. These techniques also throw light on the size of forces, although there can be a lack of agreement on the quantity of calories required and the degree to which that varied with climate, task and age; as also with horses. The extent to which soldiers could supplement their rations was also an issue.[30]

Logistics emerges as an important aspect of all military activity, including those such as the Crusades not popularly understood in that light.[31] The logistical imperatives of warfare in pre-modern times are discussed by contemporaries, including Vegetius, who emphasised the need to prepare against the threat of food shortages,[32] Sunzi, Xenophon and Froissart; although not with the detail later shown, for example by Louis Dupré d'Aulnay, the *Directeur Général des Vivres* (Director-General of Supplies) for the French army in his *Traité general des subsistences militaires, qui comprend la fourniture du pain de munition, des fourages et de la viande aux armées et aux troupes de garrisons* (1744).

Turning to Western nineteenth-century thought, the idea of progress as measured in, and by, social and economic development was readily applied to military activity, not least as the idea lent itself to contemporary interest in scientific formulation and application. Darwinism was part of the mix, as evolutionary concepts provided concerns and metaphors, notably what was to be termed functionalism, in the shape of serving goals necessary for survival and therefore strength. Rational choice and instrumental behaviour were seen as at play, from biological preference to economic and political practice. The approach to social science systematised by the German sociologist Max Weber (1864–1920) was taken into American thought by the sociologist Talcott Parsons (1902–79), an active exponent and defender of modernity, or, at least, his view of modernity, a point that was all too common and that remains highly pertinent.

In the 1960s, modernisation was regarded as a form of global New Deal, able to create a new world order.[33] Cultures that followed, or appeared to follow, a different path were presented as redundant, not least because conceiving of, and engineering, structural change was regarded as a route to success. Military history was deployed accordingly:

> the Muslim states … could no longer meet and defeat the expanding repertory of innovations developed by their Christian adversaries, because the Westernisation of war also required replication of the economic and social structures and infrastructures, in particular the machinery of resource-mobilisation and modern finance, on which the new techniques depended.[34]

Logistics appeared a key instance of this progressivist modernity, one that could readily be quantified and thus have proven effectiveness.

Moreover, the mathematically-based analysis of Operational Research, which was widely applied in and after the Second World War, and of management theory, both came to play a more pronounced part in Western thought about logistics. The use of analysts, such as those of the American RAND Corporation, founded in 1948 to provide analysis to American armed forces, was a major element.[35] Their methods encouraged mathematical modelling, for which logistics offered much. This was taken further forward in the 1980s when the entire American approach to 'preparation' and supply was rethought under the label of 'sustainability,' a concept that also affected other militaries, notably Britain. The long wars the Americans fought in Vietnam, Iraq and Afghanistan helped refine their logistical practices and concepts, although without matching success in understanding and thwarting those of their opponents. Again, there was the point about capability being unable to dictate an outcome. The study of the logistics of America's opponents was, and remains, less well-developed, although impressive work does exist.[36]

Taking a broader approach, the relationship between military logistics and both the more general nature of the state and that of organisations as a whole is highly pertinent, and it provides a context for assessing the subject in terms of the more technical details of movement. In terms of this context, the relationship between the state and organisational development attracts attention, with state organisations linked to 'interest aggregation', both domestic and international, and both military and non-military. This is an approach in which logistics becomes an aspect of the more general issue of military, political and social transaction costs, costs which were inherently highly dynamic.[37]

Phrased differently, innovations were usually a matter of modifying established institutions and practices in part by creating a new 'interest aggregation', while, generally, there was an interplay of central administration, local governance, and entrepreneurial groups,[38] all given immediacy and energy by the facts and pressures of war. Although innovative, technologies were usually adopted with a degree of adaptation that reflected established conceptions and institutions of power and social relations.[39] At the same time, precisely because 'the state' was not a bureaucratic entity but, instead, a sphere of multiple differences and drives, the processes and adoption of adaptation were related to these divisions.[40]

Moreover, the pressures of war were part of the continual uncertainty, in both wartime and peace, over resources that stemmed from harvest and disease variations, none of which were new. Thus, bubonic plague in mid-sixth century Byzantium greatly affected the resources available for sustaining campaigning in Italy and Spain, and thereby recreating much of the Roman empire. Epidemic diseases in armies and navies at war, and medical services and responses, are both an aspect of logistics and a separate matter. A badly fed army or navy was more vulnerable to disease.

Aside from issues in the standard approach to logistics, not least as far as the modern world was concerned, there has frequently been a failure to feed into the general account the perceptions gained from excellent research in the Ancient world and the Middle Ages, notably regarding the logistical support of the Crusader States and from the mid-thirteenth century the Mamluks, including the impressive fleets that supported them.[41] Such research can subvert standard chronologies of modernisation. Moreover, the range of states with very different technologies, for example the far-flung Inca empire in South America in the fifteenth century, with its lack of horses and oxen (llamas were not an adequate substitute), suggests an unpicking of the automatic assumption that modern societies are necessarily more effective.

Of course, the scale of activity is very different in the modern world, where there is a forecast population for 2020 of 7.8 billion, with 8.2 billion in 2025, and where the emphasis for logistics on range in capability and rapidity in action is unprecedented. Thus, in response to the Covid-19 pandemic in 2020–1, major organisations had to transform their businesses literally overnight as their supply chains, both international and domestic, altered in all respects. Many, but not all, had the inbuilt capability to transform, adapt, direct and manage such change from within existing resources. At the same time, extensive military assistance with logistics during the crisis, for example with the delivery of medical supplies in Britain, was another instance of the long-established cross-fertilisation of civilian and military expertise, business and the military, seen for example in the extensive American use of the latter in building railways during the nineteenth century. Military expertise has also played a key role in large-scale roadbuilding projects.

This usage was particularly apparent in the professionalization of engineering in the nineteenth century, in the extensive use of private companies, notably in the United States, in the Second World War, and in later interest in logistics as a form of efficient management shared by military and business alike, one in which each could learn from the other, with consultants providing advice accordingly. The American approach to force generation in both world wars was based on business models. As a result, publishers who had already secured strategy and leadership for their list added logistics, as when the Harvard Business School Press in 1992 published William Pagonis' *Moving Mountains: Lessons in Leadership and Logistics from the Gulf War*, a reference to the American success of the previous year.

The management of complexity is terribly difficult today, but that also throws light on the situation in earlier societies, when there were also major difficulties. The risk element then was greatly complicated by the frictions of distance and by the difficulties of countering unpredictabilities in travel, weather and disease, unpredictabilities against which defence mechanisms were generally weaker than today. This brings up the issue of comparative tolerances toward difficulties across time and geographically, and tolerances in terms of both functionality and attitudes. An aspect of such tolerance was success in improvisation, for, like most military activity, logistics is a matter of the interaction of pre-existing practices, even structures, and improvisation.

Comparative tolerance owes a lot to particular circumstances. Thus, the logistical support of the Viet Minh forces during the siege of Dien Bien Phu in north-west Vietnam in 1954 was a masterpiece of strategic, operational and tactical planning, and was the battle-winning factor which forced the French to negotiate. Yet, aside from Viet Minh effectiveness, the smaller French force was both very vulnerable and heavily dependent on the precarious nature of aerial supply.

The subsequent Vietnam War also posed questions about the respective capabilities of different logistical systems. The Americans had superb power projection and also employed counter-logistics, notably bombing the Ho Chi Minh Trail, which was used to send supplies from North Vietnam via Laos and Cambodia to South Vietnam, and, in 1970, attacking the Cambodian bases of the North Vietnamese, the latter possibly hitting stores to an extent that limited operations.[42] Yet, although taking serious casualties, the Viet Cong and North Vietnamese logistical bases and systems proved resilient and far more so than anticipated by American planners who had extrapolated onto their opponents their views of the role of Allied bombing in the Second World War, and of the operation of their own military system, particularly its vulnerability to disruption.

Comparative tolerances are also important for irregular warfare, a type of warfare that has to be hardwired into any analysis of logistics. It is both a valid form of conflict (and at tactical, operational and strategic levels), with its own logistical contexts and requirements, and one that should not be treated as a reversionary proposition in military history, as if one of the shadows of primitive or at least earlier practice and failure. This approach can then be taken forward to note separately that most 'warriors' on land today are not in regular armies, but in a range of organisations with different purposes: from neighbourhood protection to criminality, and the latter may well see more combat than most members of regular forces. There is an overlap here as 'warriors', like regular forces, need to be paid, and a failure of that essential supply results in low morale and desertion. For example, a lack of pay led to large-scale desertion from the Portuguese army in 1705 during the War of the Spanish Succession so that it achieved little;[43] so also with criminal gangs. In July 2020, a Mexican drug cartel, that of the Jalisco New Generation Cartel in the state of Jalisco, boasted of its military power in a video that appeared to show camouflaged vehicles with uniformed soldiers. To indicate the scale of violence, in 2019 there were 34,582 murders in Mexico.[44]

The functional approach to warfare and logistics, however conceived, always risks simplification and reductionism in both conceptualisation, methodology and practice, but it can still be instructive. The account, as deployed by Western stadial writers in the eighteenth century, such as Adam Ferguson in his *An Essay on the History of Civil Society* (1767), and Adam Smith in his *Lectures on Jurisprudence* (1763) and *The Wealth of Nations* (1776), with their theories of progress, argues that, while, in contrast to hunter and shepherd societies, the agricultural surplus and taxation base of settled agrarian societies, with their relatively large populations, permitted the development of logistical mechanisms to support permanent specialised military units, many, especially pastoral, societies lacked such units, and often had far less organised logistical systems. In war, the latter peoples, as well as others lacking a defined state system, relied on raiding their opponents, and generally sought to avoid battle; although there was also frequently endemic violence between villages, clans and tribes. As a result, 'little war', to employ a term of the period, set the dynamics of a logistical context in which specialised units and their particular requirements were not to the fore; a situation that continues to be the case. However, a focus on 'little war' or irregular conflict is not a major strand of military history, and is certainly one for which evidence is weakest, and notably so if statistical. Anthropology, therefore, can sometimes offer a better understanding of the dynamics of the system, and necessarily so if considered across much of the world.

Rather than thinking simply in terms of one or the other, regular or irregular warfare, hybrid systems have been the norm. They can be seen in particular with nomadic, semi-nomadic or pastoral, peoples who became imperial, notably the Mongols, Ottomans and Manchu, and also, differently, with those offering a full-spectrum military system encompassing a marked variety of tasks such as imperial Rome. Linked to this situation comes a particular aspect of assessing circumstances and change, as the conclusion of a detailed study of Roman logistics is more generally valid:

> One can see the Roman army as possessing an elaborate and permanent logistical infrastructure or as dealing with logistics on an *ad hoc* basis. While each of these models reflects certain elements of the Romans' logistical system, neither defines its essence. Roman logistics developed slowly over the course of time and it cannot be characterised in a single way.[45]

The cause and course of such variations are not only chronological, but also at any particular moment and reflect exogenous as well as inherent characteristics, the two often interacting in task-based, mission-focused, and environment-development, requirements.

Logistical difficulties tended to rise with the number of troops and complexity of equipment that had to be supported but, in turn, that rise and these difficulties have to be considered in terms of resource availability. Food-surplus and food-deficit areas existed across history, with implications not only for the ability to wage war but also for the relevant logistical organisation and historical evidence. Thus, in China and India, due to the large-scale demographic and agrarian resources that were available, most of the activities related to feeding soldiers and animals were outsourced to local village authorities and merchants. As a consequence, there could be a lack of state-generated documentation. This, however, did not equate with a lack of capability. In the late seventeenth century, the Mughals in India could dispatch a field force of 50,000 men from Agra to Chittagong, the same distance as from London to St Petersburg, but without needing to establish magazines to support the army. This was not because the Mughals were backward, but, rather, that there was no need to establish such magazines, because in India there were specialised guilds, temples, and mobile traders or *banjaras*, who supplied armies. The availability of food could lead to harvest-related campaigning, which, in turn, varied greatly including if, for example, there was more than one harvest. In the late eighteenth century, the British introduced the state-commissioned magazine system and, increasingly, discarded earlier mechanisms, although there was a significant transition period into the mid-nineteenth.

Far from the number of troops being no longer a key logistical issue, the reduction in the size of armies over the last thirty years in part reflects the logistical burden (in terms of cost) of paying and supporting large numbers. This was a particular problem with conscription, a system that meant that substantial numbers of young men had to be housed, fed, clothed, equipped, trained and paid, using up large quantities of resources but only supplying troops of relatively limited combat value and for a comparatively brief period. In part, however, this process was regarded, as in Israel, as a necessary preparation to a reservist system that delivered large numbers of trained troops for wartime mobilisation, while there was also a political rationale in terms of the 'nation under arms' serving as a means of cohesion and a constraint on unwelcome political intervention by the regular military, the latter for example a factor in France, Italy and West Germany after the Second World War.

The range of political rationale might be regarded as immaterial to the logistical techniques followed, notably the use of multiple levels of depots or stores corresponding to those of military units, as in stores for regiments, divisions and corps. However, in other respects, this range and rationale also helped set the parameters for logistical operation, especially the tasks required. Moreover, as a reminder of the need to consider a number of factors when advancing causes in any period or aspect of military history, the reduction over the last three decades in the number of troops to be supported can be attributed to the end of the Cold War in 1989, the rising individualism and hedonism of the young from the 1960s, and/or, indeed, to the greater emphasis on high-tech weaponry.

This emphasis itself poses different issues of supply: trained manpower and expensive, and specialised, spare parts, might appear to be the obvious ones, but, in addition, there is the degree to which the very cost of units reduces the logistical burden by ensuring that fewer are used (and therefore manned, maintained and stored) or risked, whether cruise missiles or stealth bombers. The logistical implications of using high-tech weaponry is a factor across history, as it entailed issues with the availability of weapons and 'platforms,' as well as with implications for parts, repairs, ammunition, and the availability and training of skilled manpower. This has been a factor not only in the Western system(s) but also more generally.[46] Linked to this comes the cost of logistics to the military in the shape of the costliness, financially and in operational terms, of particular logistical systems and plans, and the opportunity costs involved.

Logistical burden can be assessed very differently in terms of the cost to society of the military. There is no fixed answer of what is appropriate, or even necessary, in this respect, but, in functional, cultural, ideological and political terms, this cost can be judged to be too great. That point opens up another approach to logistics, namely the counter-intuitive one that, for these reasons, being a 'weak' military system, with many frictions in terms of its need to consent, could be important, or at least relevant, to success. This might particularly be true of winning support in civil wars, for example backing for the Continental Army in America during the War of Independence (1775–83). Yet, this need for consent could also be a factor in conflicts between states, where this could be true both of the home-population and of that being invaded.

Where, in contrast, logistics was 'effective,' that might lead to a burden judged excessive, thus triggering resentment, evasion and even rebellion, as, to a degree, the British discovered in America in the 1770s, with outright rebellion breaking out in 1775. Yet, employing more pressure, the Chinese state under Manchu rule proved more effective in the First Jinchuan War (1747–9) with, as a witness noted, small boats, horses and mules seized, and blacksmiths compelled to cast new cannon.[47] Logistical burden can also be regarded as systemic, as with the French military under the *ancien régime*, the system developed in the seventeenth century helping to create the fiscal background for the political breakdown of 1789 with the onset of revolution. Thus, entire tax systems can be seen as an aspect of logistics.

The present situation is different in that the percentage of GDP absorbed by military expenditure is in many states lower than in the past, in part due to practices of social welfare, whether explicit, such as health, education and pension payments, or, as in subsidised prices, implicit. As a result, it is social welfare that is seen as threatening public finances; although the situation looks somewhat different from heavily militarised states, such as North Korea, those with civil conflict, such as Syria, or those preparing to face such conflict, such as Iraq. Moreover, large-scale gang criminality imposes a different form of logistical burden on societies, as in Honduras and Mexico. This point underlines the difficulty of separating logistics from that of the costs of supporting the military. In one respect, supply systems, with distinct organisations and particular procedures located within the military, might appear obviously different to the costs of the military as a whole. Moreover, in practice, this analysis was not only less the case in the past than that approach suggests, but also far from an adequate, let alone complete, account of the present.

Another form of conditional effectiveness rested, and continues to rest, on the credibility of a power, in the literal sense of the credit it could wield,[48] although credit needs to be understood across a range of spheres. In the summer of 1676, John Ellis, visiting William III of Orange's Dutch forces besieging the French-held fortress-city of Maastricht during the Dutch War, noted that peasants bringing in forage and wood 'as to a mart … makes provisions plentiful'.[49] This clearly meets modern expectations, at least official ones, more than the French method that September of sending out raiders, burning down houses, taking prisoners, and extorting contributions of supplies to support their garrison. The Dutch forbade their subjects to pay these contributions and, as a result of the non-payment, the French executed the hostages the following May.

A clear instance of brutal inefficiency in modern eyes. But if a state had limited credit-worthiness, and that, indeed, is a classic friction, it needed to rely on force, or the threat of force, in order to raise supplies. Was this weakness, or an inevitable aspect of the difficulties of securing consent and of the exercise of power? So also in the Second World War. The Soviet army provided its soldiers with less food than its American counterpart, in calories, nutrition and taste, but that capability gap was linked to a very different political culture. Moreover, in tactical, operational and strategic terms, the Soviets were able to treat their troops as readily disposable, not least with large-scale punishments, including executions, of those deemed as insufficiently determined to fight.

For these and other reasons, logistical capability is not easy to measure and compare, let alone assess, and 'predatory accumulation' as a method of support can be seen both with centralised direction and provision,[50] and in more decentralised forms. Transferred costs contributed to the situation in which the bulk of these burdens were borne by the peasantry, although, as rural devastation greatly inhibited the ability to pay rents, tithes and taxes, the costs were widely spread. Moreover, logistical demands pressed on societies that could not readily increase their production of food, let alone munitions, so that the low level of agrarian productivity ensured that the accumulation of reserves was difficult. Tapping the economies and fiscal systems of occupied areas, a key means of the transference of cost, enhanced the destructiveness of conflict, made it more important not to lose control of territory, and also drove up the burdens of war for states on the defensive.

Care is required before assuming a process of development and improvement to the present, in particular through a nationalisation of control over the military and over the provision of support in the late nineteenth century. The considerable extent to which the modern state itself is also a coalition of interests, and with an international supply chain that is not under control, suggests caution on this head. In the late nineteenth century, well-armed, permanent, firepower-strong, state forces were used to gain and/or retain control of much of the world. This was, and is, linked to the related perception of the governmental dimension and the logistics employed, namely the increased effectiveness and apparent potency of states able to mobilise and direct resources, and to support permanent forces. However, leaving aside the particular question of the late nineteenth century, the general modern approach to logistics, that of focusing on the state, can fail to address adequately the controverted and contingent nature of governmental strength. These elements are clear in political history, administrative processes and practice, and such branches of military life as recruitment, logistics, and control and command. These elements included the wooing of the social élites and economic interests that were willing to co-operate with military change, especially the organisation of armies around a state-directed structure, but only on their terms or, at least, with their interests receiving due attention.[51]

Unlike for stadial thinkers who linked economy, government and military, the nature of the governmental context was indeed one dynamic of change, but that of military means could be very different. A key contrast across time was that of machinery, with modern military machinery requiring more sophisticated manufacturing capability as well as more maintenance, and the latter setting up more complex requirements for the troops using them. This can be seen with the use of tanks in the twentieth century, as they had to be maintained by the operators, whereas aircraft were handled by specialists at the end of each sortie. Moreover, obsolescence, and the skills and resources necessary to avoid it and separately to cope with the consequences, could be more of an issue for advanced militaries.[52] For example, vehicles remain effective as their fuel supplies decrease, but cannot operate for a minute when the fuel runs out. In contrast, troops and animals do less well when food supplies run low but can go on operating for a while.

At the same time as the problems posed by high-specification weaponry, platforms and supplies, less expensive, more durable equipment, with lower specifications, could be more useful, the durability reducing the logistical strain. This issue can be pushed further by contrasting the logistical strain of weapons and 'anti-weaponry,' such as anti-tank and anti-aircraft guns, and, in that light, assessing the resulting capability gap, and the respective cost-benefit structures, practices and doctrines of the two sides. The respective costs, both purchase and logistical, of an aircraft and a surface-to-air missile are highly instructive. The Kipling quote brings this point out for earlier conflict on the imperial frontier in the late nineteenth century.

So also with relative sophistication in finance. On the one hand, the role of money in resource availability and use ensured that the international diplomacy that could produce subsidies and loans was frequently crucial, but also gave an often decisive advantage in war to those states whose imperial and international trade systems were most developed. The larger the scale of pre-war production of supplies of all kinds, and of the internal and international trade systems that moved them, the stronger was that state. However, on the other hand, administrative sophistication, in logistics as in other facets, did not suffice for victory, as the Chinese discovered with their (eventual) total defeats at Mongol and Manchu hands in the thirteenth and seventeenth centuries.

Separately there is also the problem of extrapolation in terms of some technologically-based and costly supposed best-practice.[53] In practice, the degree of organisation and cost required to create and support a large, permanent, long-range navy, or large, permanent armies, was not, and is not, required to maintain military forces fit for purpose across most of the world.

For all, a key factor was a grasping of what was possible, and, in logistical terms, an appreciation of the effective limits of military power and of the degree to which the allocation of resources involved prioritisation or tasking. Strategy and logistics were thereby closely related, although the degree to which the latter informed the former varied greatly.[54] In considering logistical possibilities, a central element was that of continuity alongside the headline of technological change that tends to engage attention.

One central element of continuity was that of the annual variation of the weather and harvest. Thus, in 1075, the (Holy Roman) Emperor Henry IV invaded Saxony, only to suffer because the crops were not ripe, and, as a result, invaded anew, and more successfully, that October.[55] The same problem affected Maurice of Nassau, the Dutch commander, when he invaded Brabant in June 1602. The harvest and weather continue to be significant, although in some contexts more than others, for example, determining the timing of the American invasion of Iraq in 2003. American forces not needing to eat local food is an aspect of war as also waged with missiles, drones and AI. Yet, however important for some militaries, such confrontation and warfare does not define the full range and experience of warfare today, and will not do so in the future. Success in logistics and success in war is dependent upon one's ability to do what is necessary and appropriate in one's context: geographical, economic, political, technological and military.

Chapter 1

To the Fall of Constantinople, 1453

Far from being a backward and inconsequential prelude, most of world history is included in an introductory chapter that closes with Sultan Mehmed II's capture in 1453 of Constantinople, the capital of the Eastern Roman Empire. Shaping such a period risks providing a misleading account through inclusion, omission or both. This problem is accentuated because logistics involves a range of characteristics and interacts with social patterns and developments, political systems and goals, and ethical norms. It is all-too-easy to focus on the technology, as in from 'human porter to jet transport', and to underplay the extent to which, although both significant and one of the causal factors in change, the potential of the technology and techniques of supply also owe much to social, political and ethical factors. In short, logistics is an aspect of the culture(s) of war.

The basic context of logistics was that of the immediate locality and, for much of human history, the hunting of animals and raiding of other human groups were the prime means of violence, a situation that in some contexts has persisted to the present.[1] The British explorer John Speke, in his *Journal of the Discovery of the Source of the Nile* (1863), recorded the absence in parts of sub-Saharan Africa of any sharp distinction between raiding and hunting, which was partly due to the fact that non-tribal members were not viewed as full persons.

Hunter-gatherer societies would have taken food with them in such activities, but the situation changed with the development of agriculture and the related increase in population, permanent settlements, and the processing and storage of food. Linked to this came the transition from a warrior society to a soldier one, and this entailed a greater scale of activity and had implications for logistical requirements.[2] In turn, the rhythm of campaigning owed much to the harvest, with soldiers in Japan, for example, keen to return home in order to plant and harvest rice, while, supervising their agricultural lands, the officers wished to oversee the process. At the same time, rice was also purchased. There was a similar annual pattern to campaigning in other areas, as in India, where campaigning followed the autumn harvest. This was also when the roads dried out and river levels fell, after the end of the monsoon rains, while temperatures became more bearable. In turn, the fighting season ended before the harvesting of the winter crops and sowing the summer ones.[3] In Ethiopia, campaigning was after the harvest: from September to January. In southern England, in the late summer of 1066, part of King Harold's army, then preparing to resist invasion by Duke William of Normandy, had to disband for the harvest. These were instances of a key element in the interrelationship of the logistics of war with that of agriculture, one seen with food and animal availability and with recruitment patterns, and in a focus on younger and unmarried men.

In the shape of granaries, food storage was the most significant logistical development in human history, for the movement of supplies is necessarily secondary to their accumulation, preservation, and retention. Granaries were used by the Assyrian Empire during its campaigns in Mesopotamia in the ninth and eighth centuries BCE.

In turn, the counter-logistics of the devastation of crops and destruction of food stores was widely used in Antiquity, for example by Athens against Sparta during the Peloponnesian War, and by Alexander the Great against the Triballians.[4] The other key area of development was 'value storage' in the form of coinage, as it obviated some of the need for both storage and movement of supplies by a centralised bureaucracy, and substituted, instead, the more organic capabilities of networks. Coinage, like bureaucracy, were aspects of governance that relied on state formation, and these were key interacting developments for logistics, ones that occurred at different periods. Coinage made the hire of mercenary soldiers and the purchase of supplies far easier.

As with coins, another sphere of material change was wheeled transport which provided greater opportunities for logistical support, enabling armies, such as those of Classical Greece, to move food on carts, judging by the comments of Herodotus on the battle of Plataea (479 BCE). Animals such as sheep, food on the hoof, would also accompany the force. As much as possible was seized *in situ* (on the spot) which explains why the Greeks (and others) tended to fight at harvest time.

Maritime operations were more complex and could entail the use of merchants in order to hire, equip and supply ships. The value of maritime transport was seen by Alexander the Great's greatest logistical disaster – the march across the Gedrosian Desert after he failed to link up with his fleet and, conversely, with the importance of ships on the Nile to Egyptian logistics[5] as well as with William the Conqueror's invasion of Scotland in 1072 which was supplied by a fleet sailing up the coast with the army.

On land and at sea, monetarisation was part of the logistical process, as it provided a way to commodify and move resources, not least for the benefit of armed forces. Thus, in the Achaemenid Persian empire (550–330 BCE), land was granted in return for military service, the land graded as horse-land, bow-land and chariot-land according to what had to be provided, with the information recorded in a census maintained by army scribes. When personal service was not required, a tax had to be paid in silver, which gave the government the ability to move resources more easily, the money being used in part to buy mercenaries.

The sophistication of that empire did not prevent its overthrow by Alexander the Great whose army was reliant on raising supplies from the areas it invaded as a key aspect of the process and role of war in the redistribution of goods from food to people (as slaves) and other forms of loot. His well-planned logistical system depended on able subordinates.[6]

The Punic Wars between Rome and Carthage (264–41, 218–01, 149–46 BCE) saw a major deployment of forces by the two combatants over the western and central Mediterranean. In the first war, the rival powers supported troops in Sicily from Italy and Africa, while in 256–55 BCE, the Romans established an army in Africa, although, after initial success, it was defeated. In the second war, the Carthaginians deployed an army under Hannibal to Italy where local allies were crucial for logistics and manpower, as was the related policy of despoiling the farmlands of Rome's allies. In turn, 'Quintus Fabius Maximus Verrucosus,' the 'Delayer,' made it difficult for Hannibal to obtain supplies in Italy and attacked his supply lines. Scorched earth policies were matched by attacking Hannibal's foragers. Rome also used naval pressure and, eventually, the dispatch of an army to stop the movement of supplies from Africa.

This was a period in which what was later to be termed the operational dimension of war was already developed, and that despite the misleading tendency of regarding Napoleon as beginning operational art.[7] Logistics has always had the tactical side but as the scale of war increased so the operational dimension became significant.

As a reminder of the full range of logistical means, Fabius, according to Plutarch, sought to win the support of the Gods by sacrificing a large number of farm animals to propitiate them. This was an element of the role of divine support, even sanctification, as an aspect of competitive validation and prestige in warfare,[8] an element that remained significant across most of history. The logistics of such support does not tend to engage attention but was important. There was a demand for religious support, both on campaign and in the home base.

The Punic Wars were relatively unusual for Rome as naval logistics were crucial to them. More commonly, for the Romans, as for the Persians, roads, along which supplies as well as troops could move, were more closely linked to campaigning, as with the stone bridge built across the Danube to support operations in Dacia (modern Romania) in 101–6 CE. Moreover, empires maintained their roads, and roads enabled the use of wheeled vehicles which were far more efficient than pack animals in terms of the volume carried per horse. The Roman roads were built deliberately straight to facilitate faster journey times for moving troops and supplies, the two combining in the rations and equipment (weapons, armour, cooking gear, bedrolls) carried by soldiers, and to make it easier to handle wagons. Cuttings, embankments and bridges helped counter the impact of terrain. These roads were to continue to be important in post-Roman times. The Chinese were also great road builders, as was the Marya Empire of South Asia in the third century BCE. It constructed roads connecting Bengal with Eastern Afghanistan, and Patna (their capital) with South India. The Suri and the Mughal empires and later the British built upon this network.

However, the friction of distance encouraged the stationing of Roman legions for long periods in particular locations.[9] The Vindolanda Tablets provide information on the Roman garrisons on Hadrian's Wall and their supply. Each legion had a commissariat responsible for supplies. This had to interact with urban and provincial governments as well as with estate owners and merchants. These were the sources of the wagons and mules that were employed to move supplies, as the armies lacked sufficient numbers of them. Moreover, taxpayers moved food and other goods provided as taxes to warehouses.[10] River and sea transport were much cheaper, easier and quicker for moving bulk goods, but had to be supplemented by road links for the legions on the frontier. Maritime logistics, however, played a role underpinning particular campaigns. River transport was also very important in China.

Grain supplies were more generally necessary for the articulation of the Roman empire, with movements by sea from Egypt, Sicily and Tunisia to the city of Rome. It was less expensive to transport a ton of grain to Rome from Egypt by sea than from fifty miles away in Italy by land. Movement by sea was also far faster because the cart animals, in the absence of fodder, would eat all that they hauled, in about five days. The supply to the army was an aspect of this, and there was probably an overlap in the systems employed, notably private contractors. Rather than there being a system that could be readily rolled out in all areas on an isotrophic surface, with an infrastructure of roads and bases created accordingly, there was an interaction between this template and the very varied geographical and climatic constraints of the empire. There could be failures, as in 49 BCE when the price of grain in Caesar's camp in Spain rose to a great height. Sails offered a greater range than galleys and notably so if helped by regular wind patterns, as with monsoon winds in the Indian Ocean.

Late Roman Empire logistics were adequate to good, the occasional exceptions being some, but not all, campaigns outside the Empire, notably the Emperor Julian II's campaign against the Parthians in which he advanced to Ctesiphon in modern Iraq in 363, an advance well supplied by a river fleet on the River Euphrates until the siege failed, following which he unsuccessfully tried to return to Roman territory via a different route after burning his fleet. Paying the troops could be a bigger problem than feeding them, especially after the third century CE when the quality of Roman coinage declined and inflation set in. The attempt to substitute pay in kind for coins did not work. Legionary forts were largely self-sufficient, with staples grown or acquired locally, although some long-distance trade, for example of olive oil from Spain, supplemented what was locally available. It has been argued that the *mansiones* recorded on the Antonine Itinerary and the Peutinger Table functioned as military supply storehouses as well as staging posts along the road system. Things became more complicated after 395 when there were two Roman empires,

West and East, and the situation in the former speedily became chaotic, with the last Emperor deposed in 476, whereas the last in the East was killed in 1453 when his capital, Constantinople, was captured.

The need for information about threats in what increasingly for the Romans became a threat-response strategy helped ensure that information flow was a mechanism or element of logistics. Rapid, effective information flow was possible in pre-industrial times, as with the Mongol *dak* system, but the telegraph in the nineteenth century did make for a significant change. Like the Chinese, the Romans also benefited from the standardisation of weapons production. In China, Qin crossbowmen carried spares which allowed them to repair and replace parts on the battlefield.[11] Such flexibility helped Qin forces gain an edge in the final stage of the Waring States period, with their wars of unification (230–221 BCE), although other factors were more significant, notably opportunistic sequential alliance-making and war-making with the other states.

The impressive, far-flung character of Roman logistics was matched in China by the Han dynasty (206 BCE–220 CE). The support of large forces on the defensive posed a different logistical challenge to advancing beyond the bounds of current rule, because supplies could be a major problem. It was necessary if advancing into areas lacking settled agriculture to take supplies, as seen with the Romans, Qin (221–206 BCE) and Han China, and the Maryan empire which also appears to have relied not on seizing food and forage as it advanced, but, instead, on purchasing it or bringing supplies. This method, however, still required victory. Thus, Emperor Wu, the Han 'Martial Emperor' (r. 141–87 BCE), found supplies an issue in attacking the Xiongnu confederation of nomadic tribes in modern Mongolia. The anticipated logistical equation broke down as the Xiongnu, who preferred raiding, proved difficult to engage, a frequent problem in steppe warfare, but also in naval conflict. Once a sedentary army moved beyond agrarian land, it could no longer forage and had to cart its supplies, but did so subject to severe logistical limits.

As a result of the attitudes of Wu's successor, Zhao (r. 87–74 BCE), as well as of logistical factors, there was a switch by China to a defensive strategy, which was easier to maintain. At the same time, there were logistical cross-currents, as in the Chinese need to look further afield in order to obtain the horses that appeared necessary so as to resist the cavalry of steppe people. Thus, the Han sought to strengthen China against the Xiongnu, in part to gain both allies and horses, by extending their military power in Central Asia in 101 BCE. Similarly, founded by invading Turkic peoples, the Northern Wei dynasty (439–534 CE), benefited from their control of the steppe, which ensured plentiful horses.

In contrast, armies from further south in China had access to fewer horses and therefore lacked the mobility and offensive shock power of the northerners, which enabled the Sui dynasty to conquer the south in 588–9.[12] It had to adapt to (and adopt) the riverine-based maritime supply systems of the south. It was the greater ease of northerners moving onto the water than of southerners moving onto horses that largely created the north-south military gradient of China; although there was a use by the Song (960–127 CE) and Ming (1368–1644 CE), of flooding. Flooding was to be used in 1938 by the Nationalists against the Japanese, large regions in order to lessen the mobility of mounted steppe nomads.

At the same time, cavalry forced to the fore a need for adequate pasture, a factor accentuated when large forces were deployed. Cavalry were at their best at the start of any campaign and then progressively declined as problems in supply meant that horses lost condition and performance, and eventually starved unless rested. The resulting issues varied depending on the natural environment. For example, facing supply problems in the summer in the Near East when it does not rain, the Mongols could not deploy the overwhelming force necessary to offset Mamluk fighting advantages in the thirteenth century. More generally, European commentators were apt to focus on the negative consequences of winter for supplies and campaigning, but in dry, hot climates, including in parts of southern Europe, forage and water became a particular problem in high summer. Any uncertainty about the resumption of rain affected the prospects for autumn campaigning. Summer heat was also a problem in the southern United States, as in South Carolina and Georgia in the War of Independence (1775–83).

The degree of organisation in the military systems of the half-millennium after the fall of the Western Roman and Han empires is a matter that is largely unclear. The numbers involved are uncertain as are the details of preparing for, executing, or mopping up, the recorded episodes of conflict. The locations and dates of battles are themselves open to discussion, as is the relative importance of battle as organised by leaders, as opposed to, first, other forms of planned conflict,[13] secondly, spontaneous violence between social groups that did not become part of memory or record, and, thirdly, assimilation and co-existence leading to no conflict at all.[14] Yet, although the sources and analysis employed to argue for continuity from the Roman period and for a measure of sophistication have been criticised,[15] there was the ability to deploy forces and campaign, and, accordingly, to develop and sustain systems to produce and sustain armies.[16] Moreover, assembling significant forces for siege warfare, as in tenth-century Germany, assumed a logistical burden that could be met,[17] although on these and other occasions there could be issues over theft of food from local farmers, leading to resistance.[18]

There can be a tendency to regard those who assaulted settled states as 'barbarian', but this underplays the organisation of their military systems,[19] or, conversely, overplays the organisation of the settled states' military systems, and certainly underplays overlaps and interaction. There was a predatory character to these assaults (as there had been with those of the settled states[20]), notably when advancing and attacking in response to poor harvests, but that does not mean that organisation was lacking. Viking forces focused on sites with plentiful supplies, such as monasteries, and established camps or bases where they could store goods to use as supplies. These helped them deploy significant forces for major operations, such as the unsuccessful siege of Paris in 885–6; although the Vikings also benefited because significant forces for them were probably only a logistical burden of 3–4,000.

Mongol logistical support was highly effective because, in the early stages of imperial conquest, it reflected and mirrored the normal nomadic state. Indeed, the integrated logistical support inherent in the armies from the central steppes was unsurpassed at the time.

In contrast, steppe armies and societies were built to live off non-agricultural land: the horses ate the grass, while the soldiers, if necessary, drank the horses' blood and milk. As nomadic herders, they were used to moving continuously, and to carrying the means of survival with them. Their armies were accompanied by herds of horses, sheep and goats. Steppe animals were all hardy and well adapted to moving long distances with heavy loads over harsh and inhospitable terrain. The culture and lifestyle of steppe people addressed all the essential logistical requirements for large-scale mobile warfare: effective transport, effective food supply, effective maintenance of their transport and even equipment resupply, as their composite bows were made from the bones and sinews of their animals. Their collapsible accommodation could be quickly dismantled and just as quickly re-erected. The hardiness of warriors bred on the steppes also enabled operations to be carried out in winter, which more sedentary and urban opponents considered impossible, essentially because of the logistical problems it entailed. The campaign against the northern Russian princes was launched in December 1237 and only stopped with the spring thaw. Different methods, however, had to be pursued by the Mongols as tasks changed, not least the lengthy sieges required for the conquest of China in the thirteenth century. A density of resource was then at issue, and had to be deployed accordingly. Mongol dependence on horses meant that where the environment did not produce a sustained source of fodder there were limitations, notably repeatedly during the thirteenth century, in Syria in conflict with the Mamluks, but also in the Dinaric Alps in the early 1240s. At the same time, adaptabilities, however circumscribed, indicated the logistical flexibility of Mongol warmaking.

This flexibility was true of Central Eurasian nomads in general, as with Turkic peoples and leaders, notably Timur (otherwise know as Tamerlaine, 1336-1405), who was more successful than his impressive Ottoman rival. Timur's campaigns were characterised by careful planning, including thorough reconnaissance. Supplies were raised on the march, but, rather than simple devastation, efforts were made to use existing structures by levying tribute on the basis of lists of businesses and tax registers. In captured cities, there was a systematic attempt to seize goods, rather than disorganised plunder. Indeed, Timur, who, unlike Genghis Khan, was a child of the city not a steppe nomad, preferred to persuade cities to surrender and then pay ransom; he only stormed them when this failed.

The horse was not dominant across all of Asia. Moreover, although it could serve to convey troops and basic logistics in the shape of weaponry (including arrows) and immediate rations, that did not necessarily mean that horses were used as draught animals. They were not so used in China and India. In the latter, bullocks were important, not least in moving the supplies sold by the *banjaras*, mobile traders who moved with armies. Camels were important for Islamic expansion, and the two-humped Bactrian camels were an integral part of campaigns in Central Eurasia until the early nineteenth century. In South and South-East Asia, elephants were an important logistical vehicle. The absence of roads and navigable rivers, plus the availability of green trees and heavy rainfall, enabled elephants to operate in large numbers in these regions. This was not the case in sub-Saharan Africa, where the elephants were less malleable. Instead, where lakes were to the fore, as they were in Central Africa but not India, canoes were vital.

In those parts of medieval Europe where it developed, feudalism was in effect a logistical system, with the granting of landed estates in return for the provision of military service. This reciprocal dynamic, one of the ways in which governmental and logistical tasks were decentralised in the face of poor communications and inadequate monetarisation, was also seen in Christian practice, with redemption for sins in return for service. The two elements combined in the crusades; but the degree of control over crusaders varied greatly, not least given the degree to which the latter sought to follow their own priorities and make use of their own resources.[21] The pattern of logistics can sometimes be understood from the crusades. Crusader armies on the first two Crusades (1096–9, 1147–50) advanced, then stopped for resupply, either by ravaging or buying, then moved on for a few days.

The distance travelled by Crusader forces ensured that the campaigns posed unprecedented demands for Western Christendom, a situation that was greatly exacerbated by the difficulty in meeting the major burdens they posed from the particular physical and agrarian environments they traversed. As a result, the provision of supplies to the crusaders by Byzantine authorities was of great consequence, and this point underlined the political significance of relations with Byzantium (the Eastern Roman Empire). So also with maritime supplies and the Italian states, notably Genoa, whose ships brought timber for the siege machines besieging Jerusalem in 1099, and also Venice.[22] Whichever the logistical route, bases along it were significant and could help shape campaign objectives, as with the conquest of Cyprus by Richard I of England in 1191. The parameters of supply could be altered by technological means, notably the use of galleys, but there were no fundamental changes. Galley combat itself required extra exertion by the rowers and greater need for rehydration. A mastery of the logistics of water was important to naval success, as in the Mediterranean by the fleet of Charles of Anjou between 1282 and 1304. Sails could help conserve energy and water.

Within the composite armies of Western Christendom, supply was down to the individual or the particular group within the army. This might be subject to an overall policy, of plunder or do not plunder, and sometimes the ultimate commander took responsibility. Charlemagne, King of the Franks (r. 768–814) was deeply concerned with provisions for the army, but provided, with food and wagons, directly only for his followers. He ordered his major leaders to ensure that food and equipment was available, but expected this to be done by lesser men for their followers.[23]

After the Carolingian collapse in the ninth century, even that degree of responsibility declined. By the fourteenth century, government and command were more effective anew, but ravaging remained important, to feed the army and satisfy the desire for loot (as with the Albigensian Crusaders around Toulouse), to destroy crops and uproot vines, to deny food to the enemy, and to provoke them to battle or negotiation, as with English campaigning in France during the Anglo-French Hundred Years War (1337–1445). Indeed, destruction was the staple of war precisely because all societies lived close to the breadline. Supplies were obtained locally, as with the building of siege machines on site from local wood, for example for the siege of Termes in 1210 during the Albigensian Crusade in southern France. From 1209, outright seizure of resources was followed by the transfer of estates to incomers from northern France, so that they had access to the produce and revenues of the estates they took over.[24] This was not dissimilar to the Norman Conquest in England in 1066 and the subsequent suppression of rebellion. So also with the establishment of landed estates elsewhere, for example by the Spaniards in the New World.

The system of coerced contribution by the local population is particularly well-documented during the period of the Hundred Years War where the system of '*appatis*' was fundamental to the support of units both in royal employment or acting in a mercenary capacity,[25] and also in Italy after 1360 where a similar system was used by the early *condottieri* bands, and not surprisingly as many had fought in France prior to the Treaty of Brétigny of 1360. For the English, as for others, there was an interplay of strategy and supply.[26]

Although discussion of feudalism can be to the fore, in practice medieval Christian armies were made up to a great extent by paid soldiers, whether they were fighting as part of a band under a mercenary captain (*routiers*, *brabantines*, and many other terms), or whether they were part of a more conventional force. In the latter case, many of those fighting on the Albigensian Crusade were *stipendarii*, soldiers working for pay albeit under the command of one of the Crusade leaders.[27] Money, therefore, was a key component, very necessary for the widespread use of contractors to provide troops, supplies and transport, but not the sole one. Thus, Richard II of England (r. 1377–99) relied on the impressment of merchant shipping (also used by Britain in the Falklands War of 1982) to transport armies across the Channel, and then a combination of loans and taxation to meet the costs (principally wages) of the armies in the field. The armies would usually take as much in the way of provisions with them as they could at the start of a campaign, but would have to live off the land after they had been in the field for some time. On some campaigns, that caused difficulties. In Scotland in 1385, Richard II's army went hungry and had no alternative but to retreat because the Scots reacted with a scorched earth policy. Under Edward I (r. 1272–1307), this problem had been dealt with by resorting to purveyance in eastern England, and then dispatching the provisions northwards by sea, notably to Berwick. By the later fourteenth century, purveyance had largely been abandoned because it was so unpopular, and a policy of purchase substituted, which, however, was dependent on there being money in the Exchequer from taxation and loans. Purveyance, otherwise known as 'prises' (things taken), described the system of compulsory purchase by prerogative right, which was strictly speaking confined to supplying the king's own household, but which was in practice extended under Edward I to supplying the greater part of the army. Highly unpopular, it was severely limited by the Statute of Purveyance in 1362. Instead, contracts with merchants became the key means to supply expeditions, with taxation approved by Parliament the major way to pay for them.[28]

As most armies remained comparatively small, and used relatively simple equipment, it was easier to supply them from the campaign zone as long as they did not fix their position by conducting a siege. This was particularly conducive to the practice of raids rather than reliance on battle.[29] In battle, siege and, more generally, looting enemy camps (as when the forces of the Third Crusade attacked that of Saladin outside Acre in 1190), was the product of a search for supplies in the sense of loot. So also, for example, with the French attack on the English camp at Agincourt in 1415.

More generally, fortification, at whatever level, was a product of resource use and a requirement for continued provision. This represented a central goal for logistical activity, but also a means for such activity, for fortifications provided the means to store supplies. The extent to which fortress maintenance was an aspect of varied military systems was indicative of this role. Supplies were both crucial to defence and a means to attack, and their provision was significant in the interaction of the resource, organisational and fiscal aspects of logistics. Needs, however, varied with technology, including the availability of cavalry which was operationally dependent on water and fodder.

Modern criteria can be of limited applicability. The role of logistical factors in the past are often unclear, but suggestions can still be made with value, as in consideration of the problems affecting Hannibal.[30] Repeatedly, however, what is difficult to understand is how these limitations would have been perceived, and notably so in the shape of the can-do nature of military activity.

Values change in many respects. For example, soldier colonies were a classic instance of a logistical infrastructure that was effective even though different to modern systems. Roman legions on the frontiers had *prata legionis*, fields of the legion, which were assigned to them and cultivated directly by the troops or leased out on behalf of the legion. Increasingly, this extended to individual soldiers. Part of the Chinese army was usually based in military colonies, mostly near the frontiers, where soldier-farmers, provided by the government with seeds, animals and tools, produced food while also manning the fortified strongholds that the colonies contained, strongholds that, in turn, protected the storehouses and granaries. However, in the fifteenth century, this system deteriorated, in part because of the impact of Mongol attack, but largely because the officers tended to take over the colonies, using the soldiers as a labour force. Thus, the military came to approximate to the landed system found in many other states, for example much of the Ottoman empire. The change in the Chinese colonies affected military effectiveness as the colonists, while remaining militarised peasants, became more farmers than soldiers. Much of the army was raised from a hereditary soldiery, although mercenaries became more important in China during the sixteenth century, in part in response to the deterioration of the hereditary element. This was an aspect of the degree to which variation occurred within what remained discernible structural constraints.

To a degree, systems in which land was allocated in return for military service faced similar issues.[31] These encouraged substitution, notably in the form of raising money in order to hire soldiers, but also in the requisitioning of goods, such as ships. Yet, at sea, such systems rarely provided the force required, or at least without major problems.[32] On land, however, requisitioning worked as soldiers did not require the major long-term capital investment to recruit and sustain them that ships or, even more, major fortifications did.

Militia systems have greater long-term resonance than military colonies, although, from the logistical standpoint, the past ability of agrarian societies to call on local resources of food in order to support local militias ensured a very different situation to the modern idea of a National Guard that operates at a distance. Militia systems were an apparent alternative to the costs and logistical burden of regular troops, although they faced major issues in terms of skill as well as the ability to project power.

Significant continuities in warfare and logistics included environmental factors in particular. These varied from the impact of currents and winds on maritime logistics to the significance of oases of fertile land and water for campaigning on land.[33] Dependence on weather conditions was a key continuity, weather accentuating terrain factors with winter, for example, closing mountain passes and exacerbating disease.[34] Another form of continuity was provided by the range of combined-arms forces. For long, the relevant technologies available in many regions changed little, most significantly not communications, notably the inherited road system, as well as the goods moved and consumed. Thus, due to geographic channelling, Roman roads were significant. For example, the *Via Militaris* from Constantinople northwest via Serdica (Sofia) to Belgrade was of importance in Byzantine campaigning against Bulgaria, and also to the Crusader forces from the 1090s.

Another form of continuity was provided by moving meat with forces by taking animals and slaughtering them *en route*, as was done by the Romans. So also with obtaining water and most forage on campaign. Grain was usually taken with campaigning forces, which helped ensure their need for baggage trains. Grain was relatively easily transportable, had a long life, and good nutritional value, but had to be kept dry.

Alongside continuities in logistics and warfare in this period, there were also changes, including the diffusion of domestic animals. Thus, in India, the use of camels and bullocks as beasts of burden for war became more common, which helped raise the range of operation by increasing the amount of food that could be taken, although reliance on draught animals helped drive up the labour requirements of logistics. Other changes included, in the ninth century, the introduction of the shoulder rather than neck harness, and three centuries later the pivoting front axle and moveable whippletree. Together these made wagons more effective. Closing the chapter on the subject of reliance on animals is a reminder about the strong need to avoid a teleology focused on machinery or gunpowder. Indeed, in the Ancient and Medieval world, logistics were essentially muscle-powered on land, with everything relying on organic and renewable supplies, while the wind in the sails was the major addition at sea. Muscle-power is the basics of logistics, and the innovations we will encounter were simply innovations in addition. Moreover, as this chapter shows, the world of muscle-powered logistics was far from uniform.

Chapter 2

From the Fall of Constantinople to the End of Ming China, 1453–1644

Trans-oceanic empires provide a new theme in this period, one that pushes Western European powers, notably Portugal and Spain, to the fore. On land, the logistical range did not see a comparable change, for already there had been impressive long-range power projection, as with the Achaemenid (Persian) Empire (550–330 BCE), Alexander the Great, Tang China (618 CE–907 CE), the Abbasid Caliphate (750 CE–1258 CE), and the Mongols. The Ottoman (Turkish) Empire was in this sequence, and the fact that its forces used gunpowder was not a key issue in ensuring change. Despite the significance of the Ottoman Empire, the leading scholarly work for this period is a first-rate, archivally well-grounded, account of the organisational mechanism by which Spanish forces moved from Spain via Italy to the Spanish Netherlands (modern Belgium), in the period when Spain was the leading military power in the West.[1] Important in its own right, this account set a model for work on other imperial systems, notably that by Caroline Finkel of the Ottoman empire.[2] At the same time, the impressive capability highlighted by Geoffrey Parker was also accompanied by the multiple logistical weaknesses of Spain, notably, but not only, in the distribution of gunpowder.[3] In 1657–8, a major fall in the funds sent from Spain to the Spanish Netherlands led to very serious strain on the local contributions system.[4]

There is also the question of whether the focus on Spain poses the same problem as approaching modern-day warfare from the perspective of the United States. This is the case not only in looking at a cutting-edge military but also, as a related aspect, at one that specialised in long-range power projection. Each of these criteria was, and is, atypical. And, while such states deserve attention in any consideration of logistics, and notably so for the long-range power projection in which they tend to excel, leading in this book to extensive discussion of Britain and the United States, they should not serve to crowd out more typical ones. Military history suffers if it is written from the perspective of some kind of supposed paradigm power. It should be relevant for Paraguay as much as Prussia.

Part of the conceptual problem arises from the habit of relating military systems to the process of state development, such that the former are regarded as more impressive and worthy of attention if they are linked to a state following bureaucratic mechanisms. That is an approach to war that reflects the essentially functional idea of specialisation as a form of modernisation: a thesis that is seen in stadial (stages) accounts of historical sociology, such as that of Adam Smith in *The Wealth of Nations* (1776), a work which dealt with much more than just economics. In this progressive prospectus, specialisation requires a logistical system that accumulates resources from society as a whole in order to succeed. Thus, for the period covered by this chapter, as also throughout the book, we have, in accordance with this theory, as well as arising from empirical observation, a situation of significant contrasts.

In practice, whatever the degree of development, armies and navies in this period generally rested on limited, and far from predictable, systems of support, and in both peace and war. Armies and navies required manpower, pay, food and munitions, each of which posed distinctive logistical issues, and differently so on land and at sea.[5] Pay, for example, was harder to provide but, while bulky, coinage compared to modern means of paper notes or electronic transfer, was less bulky than food; although pay required more of an escort. Food and water were continual requirements: not only had the men to be fed, but also their horses and beasts of burden, such as mules and oxen. The animals had a need for large amounts of water.

As throughout the history of logistics, the overall situation hindered operational planning, but did not prevent it. Strategy, therefore, was not dictated by logistics. More specifically, transport problems helped encourage a widespread reliance on raiding, not only to deny support to opponents, but also in order to seize goods. This practice was on a continuum, via contributions from occupied 'foreign' lands, to systems of taxation or forced supplies from often unwilling 'national' or 'state' territory. The common theme was a need, in often hostile environmental circumstances, to rely on force, or the threat of force, in order to ensure consent supplies, and, often, indeed, to recruit soldiers and sailors, so that one of the major objectives of the military became its own support.

This is a theme across time and one that can be readily located in standard accounts of war and the state. This situation, while, in the abstract, compromising the ability of armies and navies to achieve operational, let alone strategic, goals, thus proving a form of friction, in practice also altered the goals as part of a more general process of adaptation, and therefore success.

The need for supplies certainly ensured that logistics was heavily politicised. Thus, the issue of how best to support forces financially helped cause political crises in England and Spain in 1640. In turn, these crises, respectively, transformed the situation as far as the war with Scotland was concerned[6] and greatly affected the situation in the Franco-Spanish War of 1635–59. As at the present, the very provision of a logistical system to support the military can be both an enabler of operations and yet also politically highly damaging, this situation indeed a key aspect of the opportunity-cost of war.

As a key instance of change, the missile weapon of preference across part of the world altered during this period from javelins, slings, and bows and arrows, to guns. The change occurred as a different process in particular places, but had major resource implications. Gunpowder, which was not reusable, could not be obtained on the march nor as a form of contributions. Given the absence of rifling on the barrels, and their windage (gap between barrel and projectile), shot could sometimes be reused, notably, if undamaged, with early cannon balls, both stone and iron, but it also could not be obtained on the march or as contributions.

Thus, the increased use of gunpowder in the West from the fifteenth century accentuated, but also altered, the logistical problem by encouraging reliance on a substance that, unlike food, water and forage, could not be obtained on campaign. In addition, there were the logistical problems and consequences, most prominently delay, produced by the need to move artillery, problems that made sieges uncertain.[7] Arms and artillery, shot and cannon balls, were all bulky and heavy, frequently getting stuck in the mud and on the uneven roads, and being difficult to haul uphill. To move them was far from easy and required labour and carts that were often in short supply. It was also difficult to keep gunpowder dry and problems were caused by moving gunpowder over poor roads. This situation encouraged the already established usage of shipping to support a land invasion, as when Charles VIII of France (r. 1483–98) invaded Italy by land in 1494, but, in a major and difficult deployment, his siege train was moved by sea from Marseille to Genoa, thus bypassing the Alps, and then by sea to La Spezia.

Developments in the fifteenth century in Europe had made cannon more effective, flexible, and easier to use. The employment of improved metal casting techniques, that owed much to the casting of church bells, and the use of copper-based alloys, bronze and brass, as well as cast iron, made cannon lighter and more reliable. Improved metal casting also allowed the introduction of trunnions that were cast as an integral part of the barrel, providing improved mobility, which was important as artillery trains were the most cumbersome and slowest part of an army. As with much of logistics, there were overlaps between production, maintenance and use. Thus, the employment of cannon was heavily reliant on their maintenance, and the availability of skilled artisans was important at every stage from manufacture and movement to use and maintenance, but, as the same individuals were generally involved, the value of distinguishing the stages can be limited.

The French artillery became more mobile under Charles VIII, with less heavy cannon mounted on solid wheels, and the artillery train reliant on smaller, more resilient carts, and made more organised. More manageable cannon and carts could be moved by horses rather than slower oxen.[8] Yet, new technology was not easy to apply, not least, in this case, because of the problems of providing sufficient equipment, including carts, as well as the difficulty posed by the roads.[9]

The most effective large state in logistical terms was China, which was able to move substantial quantities of supplies from the prime agricultural regions to the frontier areas, where large armies had to be supported in order to fulfil strategic and operational goals. This was a central theme in Chinese geopolitics and in transport links, notably the Grand Canal from the Yangtze River, the centre for rice production, to Beijing, the prime base for the key frontier zone with the steppe. These links made it easier to mobilise and support large forces, internal maritime transport and the resulting convenience in moving bulk goods being key elements.[10]

The boldness of what planning made possible for China was shown in 1547 by Zeng Xian, Governor of the northern province of Shanxi, who proposed detailed schemes to drive the Mongols from the Ordos, the arid region within the big bend of the Yellow River that was a base for Mongol pressure on China, not least Shanxi, which had been devastated as a result in 1542. The initial proposal was for an attack by 300,000 troops, but, despite the possibility of in part using water routes, the Supreme Commander, Weng Wanda, correctly argued that the problem of operating in the arid Ordos would require an enormous logistical effort that would be difficult to support. The Emperor was persuaded by the opposition, and, in part because he continued to prepare for such a campaign, Zeng was executed in 1548.[11]

Capability did not necessarily ensure the availability of supplies at the tactical cutting-edge of the operational and strategic goals. Thus, in 1449, the Yingzong emperor was captured and his army destroyed at Tumu on the return from a foolish advance toward the steppe against the Mongols. The viability of the strategy was itself problematic, but, in 1449, the key issue was the establishment of a waterless camp that was swiftly surrounded by the Mongols, and then the destruction of the Chinese army when it tried to break out. Access to fresh water was a major tactical and operational requirement,[12] as also with the total defeat of the army of the Crusader Kingdom of Jerusalem by the Ayyubin Sultan, Saladin, at Hattin in 1187. This army was kept from the Springs of Kafr Hattin and the water of Lake Tiberias, and its thirst was exacerbated by Saladin's force setting fire to the dry grass. The role of water as a source of vulnerability linked combat with hunting and controlling animals, access to water courses serving as a means of control.

The issue of the steppe was the most pressing instance of the more general Chinese problem with the border peoples they sought to dominate, an issue that has continued to the present. As with some other logistical methods, that of soldier settlement was a longstanding one that was intended not only to provide support for troops but also to deny resources to opponents.[13] The forward-deployment of troops was a key element as it made possible the rapid response necessary in order to take advantage of the campaigning season, while also permitting a multi-layered logistical response of provision of supplies to different units rather than them all operating as a massive bloc. A multi-layered logistical response was necessary not only for such offensive operations, but also because of the significant logistical strain of maintaining forces on the defensive. The latter were often particularly difficult because of the need to maintain such forces along a broad front for an indeterminate amount of time, while the attacker could focus and choose his timing, as with the Norman invasion of England in 1066.

The methods of steppe warfare continued to have considerable applicability in this period, not least in profiting from mobility and surprise. The denial of resources to opponents, notably by poisoning wells with dead animals and by setting fire to grassland and crops, was an important aspect of scorched-earth strategies, as successfully used by the Safavids of Persia against the invading Ottomans in 1514. This counter-logistics exacerbated Ottoman logistical problems, ensuring that Sultan Selim I's exploitation of his victory at Chaldiran was limited. Again, the supply of water was a key element.

In 1535, by avoiding battle, the Safavids helped ensure that logistical problems handicapped the Ottoman attack: the harsh environment of the Iranian plateau was a significant factor and scorched-earth tactics helped thwart Sultan Süleyman the Magnificent (r. 1520–66) when he advanced into Azerbaijan. Yet again, the burden of supporting large forces made it difficult to sustain the Ottoman presence. There was a seasonal dimension here too, as also on the Hungarian border with Austria, with Ottoman forces gaining control in the summer only to retire during the harsh winter.

This was instructive because the Ottomans had not only an impressive system of weapons manufacture, but also a formidable and well-articulated logistical system,[14] with the central position of Constantinople permitting the maximum use to be made of the extensive road system constructed by the Romans and Byzantines, while interior lines of communication made it easier both to apply force and to switch the point of attack. In contrast, the rival empire of the Emperor Charles V (r. 1519–56), an empire including Spain, the Low Countries, Austria and much of Italy, had France as a barrier in the centre and the Alps as a major physical obstacle, while also lacking an organisational infrastructure to match that of the Ottomans.

Chaldiran was a major instance of the mistake of risking battle when the logistical situation posed serious challenges to opponents. So also with the continued advantage of resting on the defensive and allowing opponents to exhaust resources on a siege, as with the Ottoman failure in besieging Vienna in 1529. Heavy rains that year had reduced roads to mud, ensuring that, at the end of a long supply route, the Turks lacked their heavy cannon. Campaigning at such a distance from the Ottoman base caused major logistical problems, as troops and supplies had to move for months before they could reach the sphere of operations. Alongside the movement of *matériel* from Constantinople over the Black Sea to the port of Varna and then transhipment for forward movement by water to ports on the River Danube, there was, as it were, 'the Ottoman Road' which went on the route of the old Roman road from Constantinople and Adrianople to Belgrade, and then on via Buda. Troops could march the 600 miles (965 kilometres) from Constantinople to Buda in six weeks, drawing for provisions on forty depots, while magazines were established in the conflict zone. The soldiers were well fed, mostly with biscuit (hardtack) or bread, but also regularly receiving meat, especially mutton, and the pay was not generally in arrears.[15] Centralised requisitioning and distribution was a key and effective element of the Ottoman supply system, while the use of biscuit ensured that soldiers did not have to be supplied with grain, which required grinding, nor flour, which could spoil, especially if wet.

Despite the many problems, indeed prohibitions, posed by winter ice, spring spate, and summer drought, transport of troops and supplies on rivers was also important for the Ottomans, and notably beyond Belgrade on the Danube and on the Tizsa, both of which flow north-south in Hungary. When Charlemagne, coming from the west, had campaigned against the Avars in Hungary in 791–6, he had used boats on the Danube to move supplies. Boats were also significant in moving supplies across rivers. The Ottomans established their own flotillas, but, in 1598, Ottoman campaigning in Transylvania (northwest Romania) was affected by attacks on supply ships on the Danube and Tizsa.[16] The time taken for the Ottomans to advance ensured that arrival in battle-zones was not until the summer; while winter proved a bar for operations, although, as a reminder of the need not to offer a monocausal interpretation based on logistics, that was because the troops wished to return home as much as due to logistics.[17]

As was long established, rivers were more generally significant for logistics, providing, for example, the means by which Ivan IV (the Terrible) of Muscovy (Russia) was able to move *matériel* both expeditiously and safely to support his siege of Kazan in 1552. His two winter campaigns against Kazan, the most northerly Islamic state, in 1547–8 and 1549–50, failed in large part because the Russian army had no fortified base in the region, had to leave its artillery behind as a result of heavy rains, and ended up campaigning with an exclusively cavalry army that was of no use in attacking the city. In contrast, for the third campaign, the Russians prefabricated fortress towers and wall sections near Uglich and then floated them down the River Volga on barges with troops and artillery to 25 kilometres from Kazan where the fortress of Sviiazhsk was rapidly erected. This acted as a base for siege guns and stores, and, aided by peasants conscripted to provide transport labour, the Russians advanced on Kazan and successfully besieged it.

More generally, problems in moving food and other supplies from a distance, a key issue for the Chinese on the steppe, accentuated the value of raising supplies locally, as the invading Spaniards did when they benefited from the well-developed agricultural systems of Aztec Mexico and Inca Peru. The latter means overlapped with the quest for loot and slaves, as with the campaigns of Iman Ahmah ibn Ibrahim al-Ghazi of Adal against Christian Ethiopia in the holy war he launched in the late 1520s.

In turn, a quest for loot and slaves was furthered by the range that cavalry could provide, a range that helped overcome the supply issues of being slow-moving.[18] Thus, in Africa, the *sahel* environment offered advantages, as shown by successive empires, such as the Songhai one of the late fifteenth and sixteenth centuries based at the city of Gao on the River Niger. These advantages were lacking in the jungles of coastal West Africa. Cavalry from the *sahel* (the grasslands south of the Sahara) was a form of power projection different to that of European maritime trades, and increased enslavement in Africa was a consequence of both. The European and Ottoman supply of firearms and gunpowder to Africa was a key cause and means of the slave trade, and also a consequence of logistical need for both in Africa. On the West African coast, the Asebu army of the 1620s was the first to include a corps of musketeers, their guns being supplied by the Dutch. Moreover, enslavement by African rulers, notably through slave raids, was counter-logistical in the sense of weakening opponents as well as being a means of establishing mastery, military, political, economic and psychological.

This form of conflict was also seen within polities where it was an aspect of warfare as a means to reinforce authority and acquire profit from pillage.[19] Debt-bondage was another form of slavery. Moreover, slavery could be a source of supply for armies in the form of manpower. Slaves could be a key element in the military labour market, with slave soldiers proving significant in the Islamic world. Thus, in the sixteenth century, major states, including the Ottoman empire, Morocco and Persia, made extensive use of slaves. This use of troops provided a different context for responses to the supply of food, although conditionality and dissidence were still aspects of conduct among slave soldiers, with, for example, large-scale janissary mutinies in 1621–2.[20]

In addition, the very different use of slaves, and, more generally, of coerced labour, as porters, who unlike slave soldiers were unarmed, was an aspect of the logistics of armies in sub-Saharan Africa, and elsewhere. Slaves also provided motive power in the form of rowing galleys, a source used from Antiquity until the nineteenth century. These slaves were kept unarmed, and were crucial to motive power rather than providing military manpower. However, they were an important aspect of the logistical requirements of navies.

Meanwhile, the sites of conflict were at once a shaping and a consequence of very different contexts and dynamic for logistics. Thus, at one level, sieges, a continuing but, compared to battle, frequently underrated feature of military affairs, were a logistical struggle, with each side under pressure in a more focused way than usual. This struggle brought to fruition the implications of fortified positions for campaigning. Fortresses posed a challenge for invaders: either using resources reducing them or bypassing fortresses and risking their garrisons harrying the supply lines. Charles V's invasion of Provence from Italy in 1536 failed because the French, instead of fighting (other than from ambushes), retired into fortified cities while denying food to Charles' forces which were reduced to searching for it before retreating. Similarly, the English invasion of Scotland in 1542 was stopped by a combination of supply shortages and the obstacles posed by fortifications which imposed delay and thus put further pressure on the supplies. Best-case predictions of success were, as ever, a major problem in logistical provision.[21]

The logistical pressure of sieges helped explain the conflict that went on round them, including, on the part of the defenders, both sorties and relief attacks. The role of logistical pressures led attackers to risk assaults, as with the unsuccessful Ottoman sieges of Valetta (capital of Malta) in 1565 and Vienna in 1683; but also encouraged both sides in sieges to negotiate in order to bring the siege to an agreed close, as in particular in India (all aspects of the warfare discussed by the Roman commentator Vegetius, in his *Epitoma rei militaris* in the early fifth century CE).

At the same time, the problem posed – notably if short of space, with emphasising logistical factors, as opposed to putting them into a multifaceted analysis – was illustrated in 1568 when food and water shortages claimed much of the Ottoman-Tatar force that had unsuccessfully besieged Astrakhan on its return journey to Azov. This appears clearly to show that logistics placed a limit, but a range of factors were at play, including the shortage of cannon that affected the siege, as well as the lack of viability of the initial plan for a canal between the Don and Volga rivers, and, having retaken Astrakhan from Russia, co-operation with the Uzbeks against Safavid Persia (Iran).[22] A lack of support from the Crimean Tatars, which had also been a factor in 1569, hit plans for a joint Uzbek-Ottoman attack on Astrakhan in 1587–8, and, in 1590, the Ottomans refused to send troops to help in such an attack.[23]

Thus, logistical capability has to be assessed as part of a broader range of factors, including the role of regional allies, the possibility of gaining local support, and the geopolitical contexts of possibilities and priorities. Moreover, distance in pursuing expansion was in part an expression of mental horizons of ambition and concern and, alongside the practicalities of transport, these need to be borne in mind. The realist approach very much focuses on what was achievable, and therefore on 'the optimisation of available resources' and on an understanding of what are presented as real limits.[24] However, such an assessment was less clear-cut than might be appreciated and, in practice, was mediated through personal and factional rivalries, and therefore of related assumptions. This point does not tend to arise from the standard 'so many pounds of food daily, so many men/horses' approach to logistics; although the latter approach has great value not least for helping establish the parameters of possibility.[25]

There is also the problem posed by the widespread assumption that size inherently is better, such that logistical capability was a matter of supporting large forces. This assumption is part of a top-down account of military developments that very much focuses on state activity, and, in that, of powerful states. To query this is counter-intuitive for many, but was an issue raised at the time. In 1547, Sahib Giray, the talented khan of the Crimean Tatars (r. 1532–51), advised his ally and overlord, Sultan Süleyman the Magnificent, that the large armies used by the latter against the Safavids were less useful than a reliance on smaller, more mobile, forces. Such forces, which he, of course, could provide, for most comments on military matters served a purpose, were far easier to support logistically, and thus less of a pressure on local resources, as well as less expensive, and also better able to fix and fight the Safavids who used similar forces.[26]

This advice, at once, as with so much else in historical appraisal, whether contemporary or subsequent, both objective and self-interested, was not followed. Furthermore, although the high-cost methods of the Ottomans, with the burden of support posed by large infantry forces backed by cannon,[27] and the availability of plentiful gunpowder, helped lead to territorial gains, there was no commensurate ability to support and sustain them; a pattern seen with other states across history. There were, however, instances of success for the Ottomans (and other states in similar circumstances), notably in 1578 when Lala Mustafa Pasha, one of the many Ottoman generals who does not have the fame he deserves, achieved impressive successes in the Caucasus, despite the logistical strains in the mountainous region of supplying the invading army. The Safavids were defeated at Çıldır, the submission of local rulers was obtained, the city of Tbilisi captured, and the province of Shirvan conquered. Yet, Giray was also vindicated, for, in 1579, when the Safavids in turn counter-attacked, besieging Derbent on the Caspian Sea, it was relieved after the Tatars advanced from the region of the Kuban along the north side of the Caucasus, a dramatic instance of power-projection by a mobile force that was different to the Ottoman field army. The eventual peace terms in 1590 reflected the ability to project power, Safavid limitations in sieges, and wider power-politics in the shape of Uzbek pressure on the Safavids.

The evidence of advances on the steppe suggests that Giray was not only correct for the Ottomans. Shah Abbas I of Persia (r. 1587–1629), a dynamic and largely successful warrior, was greatly affected by the matrix of constraints produced by distance, terrain, logistics, costs and a range of opponents. This was notable, after the capture of Herat in 1598, when he marched on the city of Balkh (in modern northern Afghanistan) in

1602 with 40,000 troops, including 10,000 musketeers and about 300 cannon, a larger field force than those in Western Europe at that period, although one dependant on the presence of the ruler. In the event, a lack of provisions and water caused problems that led Abbas to a humiliating retreat,[28] with most of his cannon abandoned, a frequent fate as they were the heaviest and slowest load. Logistics was certainly an issue, with the size of the Safavid military, and the inclusion, by the 1600s, of a significant infantry and artillery, notably the corps of *ghulams*, increasing the operational strain posed by its campaigning and making it more similar to the Ottoman army.

Yet, logistics also has to be located in terms of goals and tasking. If the Safavids lacked an equivalent to the logistical support system in China, the effectiveness of the latter at this stage was not only due to resources, communications, and administrative skill and experience, but also, at least in part, to the Chinese focus on the defensive which lessened the logistical burden of operations. Thus, the Ming, as brutally shown in 1449, lacked the capability for power projection onto the steppe that was to be displayed successfully from the 1690s by their Manchu successors (see chapter three) who could draw both on a steppe component to their military and on the benefits of more agrarian productivity in China. Separately, in 1602, aside from logistics, Abbas also suffered from being outmanoeuvred by the Uzbek leader, Baqi Muhammad Khan. The following year, Abbas was able to capture Tabriz from the Ottomans.

This situation calls into question the focus in so much of the scholarly literature on the establishment of large professional infantry forces, and the tendency to equate that with long-term development, indeed progress, as opposed to seeing it primarily as a response to particular circumstances. Thus, and the point remains relevant, professionalism, like logistical capability, had very different meanings in particular contexts, and, therefore, with reference to contrasting expectations. On land as at sea, the technological context helped affect the factor of distance, and notably so as far as the availability of horses was concerned.

A parallel argument to that of Giray was provided by those Western commanders who sought to use mobility and surprise in order to break free from protracted and expensive siege-centred warfare.[29] These practices in part drew on the established practice of the raid, as in Anglo-Scottish warfare, the raid being a very different (and far quicker) route than the siege to loot, benefit and control.

Cavalry were often armed with firepower, whether bows or guns, and light cannon could also be taken. Thus, there was an overlap with the deployment of cannon at sea, where ships brought mobility, while cannon certainly enhanced fighting ability. However, to carry cannon, and then to increase the number of cannon, galleys had to become bigger and stronger, which meant larger crew and, therefore, the need to carry more provisions at a time of rising food prices. The resulting logistical burden influenced the viability of galley warfare, although, as a reminder of the requirement for contextualisation, there were also changes in construction and maintenance costs, as well as the availability of inexpensive cast iron cannon of suitable quality.[30]

Alongside distance and resilient political cultures, logistics was one of the factors that combined to ensure that it was difficult for states based in Burma (Myanmar) or Siam (Thailand) to subjugate the other for any length of time, which matched the experience of the more far-apart rival states based in Persia and the Near East; although the latter could call on cavalry and the area in-between, modern Iraq, was under their control. In 1593, a Burmese attempt to reimpose control on Siam failed due to the impossibility of sustaining the war effort. The demands of the Burmese army had placed too great a burden on the economy and society, leading villagers to flee, and hitting the ability to recruit. Burmese failure was followed by a successful Siamese invasion of Burma.

The contemporaneous large-scale conflict between China and Japan arising from the Japanese invasions of Korea in 1592 and 1597 also reflected the problems facing both sides, although the Sino-Korean allies benefited from having stable land and sea lines of support from China. The Japanese, while being able to get some supplies from the homeland, were stymied by guerrilla efforts on land which restricted their chance to raise supplies, and by the Chinese and Koreans at sea which stopped them from supplying their armies on the western side of the peninsula. The Japanese also failed to extract the expected resources from the Korean countryside, mirroring their supply woes in China during the Second World War, on which see chapters 8 and 9. Unlike the Japanese, the Chinese and Koreans made use of Buddhist priests, peasant conscripts, and soldiers, to transport supplies, and enjoyed the fact that, again, unlike the Japanese, they moved through friendly terrain. At the same time, the increase in Chinese numbers ensured that, by the end of 1597, when there were probably about 75,000 Chinese troops in Korea, they were affected by grave logistical problems.

As a reminder of the difficulties of gauging capability, and the need to consider comparisons across time, the Japanese, while defeated at sea by the Koreans in 1597 and 1598, withdrew in the spring of 1599 from Korea without having been driven out – in part, not due to the war there but because, after their commander, Hideyoshi, died in September 1598, there was a focus on civil conflict within Japan. The Japanese had the manpower and agricultural resource base necessary for operations in Korea in the 1590s and, had they continued, the 1600s, but there were also resource and logistical strains for Japan in operating on mainland East Asia and notably in sustaining operations there, in the shape of logistics, beyond the first campaign. In particular, the mountainous terrain of much of Japan limited food production. This issue was also to be seen during the Russo-Japanese War of 1904–5, and then again when the Japanese invaded, first, Manchuria and, subsequently, the rest of China between 1931 and 1945. Logistical issues, however, did not prevent campaigning, again, in part, because the idea of limits was a political matter, rather than one that could be quantified and measured, and thus fixed.[31]

In the later 1590s, the Chinese also had to face rebellion by aboriginal peoples in the southwest of China. These were troublesome in military terms, in part because the rebels used guerrilla tactics and, in part, because the difficulty of raising supplies locally, combined with the distance from centres where food stores were available. Moreover, the mountainous terrain contributed to the combat and logistical problems faced.[32]

The range of force types and logistical systems apparent in Asia were also seen in Africa, where, again, there were contrasts between forces that in part reflected socio-economic differences, although more was always at issue. Thus, Idris Aloma, *mai* (ruler) of Bornu (1569–*c*.1600), an Islamic state based in the region of Lake Chad, made careful use of economic measures, attacking crops or keeping nomads from their grazing areas in order to make them submit, both of which were long-standing practices in such campaigns. Similarly, in Ethiopia, the use of firearms was restricted by the limited availability of shot and powder, a frequent problem affecting their use, and thus the diffusion of the technology. In the same way, the Cossacks, who conquered Siberia from 1581, had to use powder and shot with care, as they were difficult to replace due to the strains of a supply chain that was very long, slow and vulnerable. Ethiopia was put under pressure by the expansion northwards of the Oromo, nomadic pastoralists who made effective use of horses, which increased their mobility, helping their raiding parties invade territory and live off the land, and thus outmanoeuvring the more cumbersome Ethiopian forces.[33]

The European powers had varied logistical systems, as befitted their very differing territories, resources and commitments. The strain of conflict could lead to a measure of administrative logistical change, both in terms of administrative practice, as with the allocation of tax revenues directly to a commander,[34] and with reference to an extension of government, either in co-operation with existing authorities or more unilaterally. The quality of the organised response depended in part on the consistency of the commitment and, therefore, the character of the need. Permanent requirements ensured a constant need for logistical support. In particular, fortification systems, in the sense of fortresses that were adequately garrisoned and supplied, and able to provide more than an *ad hoc* remedy, in large part arose when a frontier zone was both relatively fixed in its location and contested, which was increasingly the case in the late sixteenth century with defences against the Ottomans.

Thus, following the Peace of Augsburg with the German Protestants in 1555, the Emperor established the Imperial War Council in 1556 and regularised the transfer of funds from the Empire (Germany) and the Habsburg dominions, and established a new administrative system for the Hungarian frontier with the Ottomans. Armouries were transformed into arsenals, while a new post, the Chief Provisions Supply Officer for Hungary, was created and, in 1577, a military conference in Vienna produced a plan for the administration of frontier fortresses. The *Militärgrenze* (military border) of the Habsburg monarchy was to help stabilise the frontier zone.[35]

Nevertheless, when war with the Ottomans resumed in 1593, the Austrian army soon collapsed due to serious logistical problems.[36] Ottoman logistics were more effective, although the permanent garrisoning of numerous border fortresses necessary to the system created financial pressures,[37] while the Ottomans also faced serious logistical issues when campaigning, as in 1598 when they unsuccessfully besieged the Transylvanian city of Várad (Hungarian; Oradea Romanian; Grosswardein, German), only for their supplies to fail, the army to be badly affected by heavy rainfall, and the troops to mutiny.[38]

At the same time, the logistical underpinning of war for Austria, and, indeed, for many, but not all, Western states,[39] remained heavily dependent on the support of the local Estates (Parliaments). Conventionally, that dependence would be regarded as a source and product of weakness. However, that approach both underplays the value of consent, and the support thereby obtained, and, separately, the extent to which there was no fixed political culture that can be employed if assuming a background within which to assess logistical capability. Furthermore, the nature of consent was amplified due to the need to consider the role in the localities of the local élite, whether or not they were part of the formal structure of the state.[40]

Consent was differently tested during civil wars which posed contrasting problems in tasking and logistical requirements to that of state-to-state conflict; although there was also an overlap in the methods employed. The logistical problems posed by civil wars, notably the weakness of existing governmental mechanisms, did not prevent solutions in the shape of new systems (and eventual victory), as was seen with the English Civil War of 1642–6. However, these new systems still had their limitations, a caveat that is not intended to suggest criticism because limitations are the nature of life. For the Royalists and the Parliamentarians in the English Civil War, there was a lack of logistical planning, and, to use a later vocabulary, staff appreciation of the force logistical requirements all too often did not happen. The two sides knew vaguely, in a finger in the air fashion, what they required and wanted, but, due to a lack of money, simply could not fund or resource it adequately, which was especially the case of arms, ammunition and powder. The situation was exacerbated, in a period before turnpike roads, which did not begin in England until the 1660s or become comprehensive until the mid-eighteenth century, by poor communications which emphasised the importance of fortified towns or strongpoints at key locations. Each side used both the magazine and requisition systems, which was in part a response to the trouble of establishing fiscal systems that would work, a task finally achieved with some difficulty by the Parliamentarians.

The main magazines were London, Hull, Plymouth, Warwick and Reading/Windsor for the Parliamentarians, Oxford, York, Berwick, Carlisle, Newark and Chester for the Royalists, and Bristol for both sides when they held it. Fortified positions protected and supplemented the system with the garrisons, such as Parliamentary Leicester, the bases for the incessant search for supplies, a process given further geographical shape by control over bridges, for example over the Rivers Trent, Thames and Severn.

As with warfare elsewhere, the Civil War saw growing demand on the localities from both sides, as the cost of the conflict rose, and resources available in the early stages were used up. As part of an intensification of control, including over loyalists, reliable and experienced commanders replaced local gentlemen in positions of local power, Sir William Russell being succeeded as Royalist Governor of Worcester in December 1643 by Sir Gilbert Gerrard. The influence on regional logistics during the Civil War of the Clubmen, a rural local-protection movement of 1644–6 against the demands of garrisons, was significant. The Clubmen sought to restrain the demands of garrisons such as Hereford, and therefore to keep troops out of their areas. Similarly, in December 1631, there was armed local resistance in the Meuse Valley (in modern Belgium) to attempts by Spanish troops to take up winter quarters. The value of such opposition to opponents helped encourage the exploitation of disaffection by the latter, as with France and Spain each exploiting rebellion in the 1640s.

In turn, the importance of obtaining supplies in the field encouraged the devastation of possible supply sources as, by both sides, in the war between Denmark and Sweden in 1563–70. The potential of such supplies also depended on whether an army was advancing toward positive victory, and thus not worth offending, or retreating having been defeated. Thus, Hugh O'Neill's Irish army marched south to Kinsale in late 1601, seizing food, but, on the retreat after defeat there by a well-resourced English force, they were treated harshly in a now denuded countryside by those they had despoiled. In contrast to England, the Irish rebels who rose in 1641 had a weak supply system which made mounting lengthy sieges difficult.[41]

There were the very specific variations in military environment, notably, across Europe and much of Asia, between well-settled agricultural zones that were usually fortified, and, on the other hand, areas of great distances and small populations where there tended to be a reliance on cavalry and the use of devastation in order to reduce opponents' fighting capability. In the former, cannon played a greater role, and there was a requirement for a relevant supply system. Thus, 10,000 cannonballs were fired by the Dutch in their successful two-month siege of Spanish-held Groningen in 1594. That creates one impression of logistical proficiency, but a different one is offered by the dispatch of Dutch forces to the provinces of Drenthe and Groningen in 1604 and 1605 in order to make them pay the money owed to the Dutch war effort. More generally, as an instance of contrast between agricultural zones, there was a shortage of grain in pastoral areas, and this made them difficult for the operations of large armies.

The limited nature of Intelligence knowledge and capability could be such that the particular issues in specific environments were barely understood. Alongside physical and military environments, their political counterpart was significant. The Dutch rebellion against Spain saw, in the 1560s and 1570s, military entrepreneurs raising forces of their own accord, but this practice worried the urban oligarchs because they feared that mercenaries would leave them to their fate when the pay was in arrears or, even worse, betray their towns to the Spaniards. To ward off this danger, the oligarchs declared that supreme command over the troops lay with the provincial States [Parliament] and the States General, and the right to appoint officers was assigned to the former. The army thus served as the military expression of the Dutch rebellion. Therefore, the circumstances of the individual struggle were important to the development of particular forces, and this process had obvious implications in terms of their logistics. So also with the Spanish army that suppressed the 1591 Aragonese revolt. Scarcely professional, it was largely recruited by the Castilian nobility from their estates, but it was totally successful.

These very different responses, in the case of the Dutch and Aragonese revolts, reflected the need to ground logistical provision on a degree of consent, for the cost of supporting the military burden of war and peacetime expenditure threatened political stability, both thereby creating new tasks for the military and making it harder for them to fulfil these tasks. The issue was affected by the length of any struggle; for repeated campaigns generally put a serious strain on both resources and popularity, as in the Ottoman empire.

However, logistical effort and costs were more supportable if the economy was growing and population rising, as on the global scale in the sixteenth century, and, far more, despite an interruption in economic growth in the 1930s and early 1940s, from the late eighteenth century on, and less so otherwise, as in the seventeenth century. War in the latter circumstances, and civil warfare in general, could encourage a recourse to more local logistical responses, which can be presented in terms of a reversion to more primitive military arrangements, and was certainly destructive, as well as serving as a warning to those who were not compliant.[42] Sexual violence or romance could be an aspect of the situation, with an increase in the percentage of illegitimate births as a result of armies passing through and also as a consequence of the presence of garrisons. Yet, to complicate matters, violence was also a consequence of the breakdown of the multifaceted links involved in billeting and sexual relationships.[43]

At the same time, the idea of reversion to a more primitive arrangement can be a less than appropriate assessment of systems that were pursued because practical and therefore fit for purpose within that key context. This can be seen with the German practice, seen in the sixteenth and early seventeenth centuries, of *Brandschatzung* (Fire-Treasure): extorting money and supplies from villages under the threat of being razed to the ground, a process overseen by a *Brandmeister*, who was also responsible for preventing looting.

So also in terms of other judgments, for example of achieving a concentration of resources and deploying them with less need to consult local élites. This process can be seen as an aspect of the relative logistical sophistication enjoyed by the Ottomans, or as a factor enjoying no particular merit because of the value of such consultation and grounding, as in much of Christian Europe. Indeed, reliance on these élites can be presented as in part an aspect of the general reliance on private entrepreneurs to organise war and on local populations to pay.

The early seventeenth century was to indicate that these practices were the most effective way for the recruitment and supply of Western European armies. Even so there was only so much that could be achieved by these methods, or indeed any ones, in the face of very poor harvests, such as those of 1630, 1649–51, 1661, 1693 and 1709. The logistical strain of the far-flung European wars of 1683–1721 was greater than that of those of 1793–1815 in large part because of the serious demographic and economic difficulties of the earlier period, whereas, in the later one, the situation, while still very grave, was different due to significant demographic and economic growth from mid-century, growth that fed through into rental income and tax yields.

There was a parallel situation at sea, with naval forces in particular reliant on mercantile support.[44] At the same time, the role of government in developing and sustaining naval strength was more necessary, because navies and the naval infrastructure were expensive to maintain, required substantial fixed, and recurrent, expenditure, and operated from a limited number of ports. The result could also be impressive. Alongside the Ottomans, who were also rapidly to rebuild an effective fleet after defeat at Lepanto in 1571, and also Venice, the Arsenal of which was a formidable facility as well as a storehouse, Spanish bureaucracy displayed agility, dedication and inventiveness in keeping the fleets supplied, with private contractors and public officials working fruitfully together. Forest legislation sought to conserve timber stocks, efforts were made to provide sailors and soldiers with nutritious food and good medical care, and severe discipline was enforced on erring fleet commanders and bureaucrats alike.[45] Other states, such as the Mamluks of Egypt and, even more, the Mughals in India, lacked the comparable logistical commitment to naval preparedness.[46]

The great costs of naval operations were food and wages, because fleets could not live on local resources as armies did, as with wayside grass for the animals. Instead, fleets had to be provisioned and watered, needs that made the size of the crew a particularly serious matter. Water was heavy and bulky, and posed major problems for a galley's ability to stay at sea because of the size of a galley's crew. Despite the possibilities offered by obtaining supplies from fish and rainwater, this issue was far more significant for long-range voyages,[47] and led to the establishment, by the Portuguese in particular, of port-bases whose primary purpose was to replenish passing fleets. British coaling stations in the late nineteenth century were in this tradition. The Portuguese bases replicated, but over a vastly greater distance, the role of the ports that were so indispensable to Mediterranean galley operations, for example Modon (Menthone) and Corone (Koroni) in Morea, 'the eyes of Venice', and bases *en route* to the Aegean, Cyprus and Crete. Portuguese sailors knew that they could replenish in safety at a series of 'way stations', such as Luanda and Mozambique, on their long voyages to and from Asia.[48] Cape Town was to do the same for the Dutch, and Mauritius and Réunion for the French.

The availability of bases, such as Suez from 1517 for the Ottomans (previously a Mamluk base), gave a major advantage to those defending areas of water, and helped focus conflict on these port-bases, as with Tunis in 1535 and 1570–4. As has remained true today, not least with the dependence on containers and unloading facilities, for modern maritime logistics, few harbours and anchorages were able to support and shelter large fleets transporting substantial numbers of troops, a situation which affected operational methods and strategic goals, as access to, or the seizure of, these nodal points was crucial.[49] This logistical capability, in turn, provided a critical political dimension. Thus, as long as the Ottomans had access to French ports, notably Toulon, through diplomatic efforts, their naval range was greatly enhanced. Facilities could be improved, as under British rule with Gibraltar which lacked plentiful fresh water, and therefore had to have its storage enhanced.

At the same time, the demands of maritime expeditions made them a logistical problem, as with the Ottoman invasion of Malta in 1565 in which major supply difficulties included that of water. In contrast, the defending Knights of St John had secured adequate ammunition before the siege. With Famagusta in Cyprus in 1570, Venetian relief vessels did break through with supplies, but the defenders, their gunpowder running down, surrendered. Naval resupply was also important, as with Swedish-held Riga's success in thwarting a major Russian siege in 1656. Such supplies were vulnerable to factors from storms to enemy action. Thus, the Ottoman campaign against distant Portuguese-controlled Hormuz in 1552 failed because of the sinking of a supply ship, leaving the troops short of munitions and food.[50] In the 1640s and 1650s, Venetian attacks in the Aegean affected the Ottoman siege of Venetian-held Candia on Crete, preventing or delaying the movement of supplies and diverting resources.

A significant difference between land and sea warfare arose from the more varied fighting specifications of the former, and, therefore, the differing logistical contexts within the same force. Cavalry or infantry, firearms or slashing and stabbing weapons, established a key matrix in this respect. Far from uniformity being necessary to success, it was combined arms capability that was more significant. In part, this capability could draw on the nature of composite states, as with the Ottoman empire,[51] but manpower and other supplies from outside the empire were also significant, as with Mughal India and the important horse trade to it from Central Asia.[52] The Mughals ran a hybrid military and logistics system, one that, like the Ottomans, looked to former Central Asian patterns of light cavalry-based activity as well as to that of the sedentary areas they had conquered. With the Mughals, local allies were very important to this hybrid pattern as also was improvisation.

The significance of money for logistics was repeatedly demonstrated in this period,[53] and in very different ways, notably the frequency of mutinies. With the Ottomans (and others), political factors could play a role, but mutinies generally arose when pay was not forthcoming and when credit with the local communities was withdrawn.[54] The prevalence of desertion was another significant aspect of logistical pressure, as, during the Thirty Years' War (1618–48), with the Swedes in the devastated countryside near Albrecht Wallenstein's defended position at the Alte Veste close to Nuremberg in 1632. More seriously, in 1624, when the logistical system of the Mughal army in the Deccan collapsed, the soldiers, denied pay and food, joined Diwan Malik Ambar, the Ahmadnagar warlord, against whom they were campaigning, because he promised them both.

Wallenstein (1583–1634) was a key military figure in the first part of the Thirty Years' War, a Bohemian military entrepreneur to whom the Emperor Ferdinand II entrusted his forces. From his estate, Wallenstein produced bread, beer, clothing, footwear and horses, all of which he charged to the Emperor. Using Bohemian resources, which had been freed up by the suppression of rebellion there and the consequent expropriation of land, and benefiting from his ability to seize resources from the areas in which he campaigned, he built up a large Imperial army. This army then supported itself by contributions and by seizing supplies in a fashion that brought great damage to Brandenburg and other principalities. Their dukes deposed for rebellion, two duchies of Mecklenburg were given to Wallenstein, whose ability to distribute the resources raised by contributions ensured control of the troops. However, the accurate sense that Wallenstein had built up an expensive army that was not under control, led Ferdinand II in 1630 to dismiss him and, after he needed to be recalled in 1632 to oppose the Swedes, to support his overthrow in 1634, during which Wallenstein was killed. Meanwhile, the costs of supporting their forces led the rival Swedes and the allied League of Heilbronn to permit units to levy contributions in 1633, which helped to undermine their campaign that year in south Germany.

Money affected the size of forces deployed as combatants, as in the French Wars of Religion from 1562 to 1598,[55] and this decline in size could be linked to a decline in effectiveness in particular conflicts and armies. However, other factors also played a role. Thus, in the latter stages of the Thirty Years' War, there was, alongside a reliance on garrisons to guard areas from which necessary contributions were raised, a resort to smaller field forces, and also to cavalry, despite the latter costing more to hire and pay.[56] This shift helped provide important opportunities for campaigning, which poses an important qualification to the usual assumption that larger armies offered greater success. Indeed, the latter stages of the war saw wide-ranging campaigning, although serious difficulties in the supply of men, money and provisions were important to their outcome. Dominating north Germany, notably, but not only, Brandenburg, from which they raised contributions as they chose, the Swedes had a logistical basis for operations to the south. For them, and others, areas seen as resource-rich attracted particular attention, as with the French interest in Swabia in the 1640s. These were contrasted with areas that were essentially just ravaged not occupied, such as Bavaria for the French and Swedes, because their armies knew they could not overwinter there. The ability of French and Swedish forces to advance far into the Empire, the Swedes to Bohemia, helped lead Ferdinand III to terms in the Peace of Westphalia in 1648. At the same time, however attractive it might be baldly to state the proposition that the search for supplies controlled strategy and operations, in practice this search did not so pre-empt other military and political factors that it could be termed as controlling; influence was more accurate.

Cavalry was similarly significant in the war between Portugal and Spain from 1640 to 1668.[57] Moreover, the consequence for logistics of the clash in military types was seen in the failure of the Russian siege of Polish-held Smolensk in 1632–4. In contrast to the delayed arrival of Russian artillery, due to poor weather and primitive roads, there was the operational flexibility the Poles enjoyed thanks to their superiority in cavalry, a flexibility and mobility that was translated into success in the battle for supplies.

Whatever the size and type of force, funding was linked to the level of operational activity, but, at the same time, whatever the funding, there had to be a response to local environmental circumstances, as when the Mughals pushed into the Himalayan territories of Baltistan and Ladakh in 1637–9 after a difficult expedition over hazardous mountain passes by 12,000 men who had to carry all their supplies with them because the route was largely barren. As a result, draught animals could not be used, prefiguring the problems British and American forces were sometimes to face with mules in Burma (Myanmar) in 1941–5.

Climate and human support were both part of the logistical equation. In mountainous Asia, campaigning was easier in the summer, but in most of South Asia, on the other hand, the winter offered relief from heat and rain. The itinerant grain merchants important to armies in India, did not operate in most of the arid regions beyond India, and, in besieging arid and distant Kandahar in Afghanistan in 1649–53, the Mughals were obliged to take their supplies with them, only to have to abandon the sieges and fall back. In contrast, in parts of India, notably Bengal and Assam, monsoon rains caused serious problems, notably making rivers uncrossable. In 1662–3, the impact of rains on the supply lines of forces campaigning in Assam caused a serious shortage of food. Moreover, thick forests ensured that woodcutters were important to clearing a path, in contrast to the pioneers making one in mountainous areas.[58] In such different contexts, force could work best if it was a focus for the clientage and local alliances that were crucial to stabilise border regions.

This was somewhat different from the idea of the state as a proto-Leviathan,[59] but the rhetoric of power tended (and tends) to be very different to the reality, often necessarily so, while the later assessments of political scientists generally fail to match the more complex realities of the time. In practice, bureaucratic processes were mediated through societies that were still dominated by the landed orders. Moreover, in pursuit of the general theme of this book, the confidence that might have been expressed half a century ago, when modernisation theory was at its height and highly influential in academe, that public provision and state organisation were necessarily better, and therefore the goal of military development in the past, can, instead, be seen better as a projection of a particular set of assumptions. It is certainly possible to draw attention to the weaknesses of public provision, not least of state structures focused on clientage, and to the weaknesses of governments. As a result, the value of private provision can be stressed. Indeed, even 'progressive' forces, such as the New Model Army of the English Civil War, still relied on such provision.[60] Far from war making the state and the state war,[61] war could break the state and certainly generally led to serious pressure on government, notably debt, inflation, debasement, and transferring costs to officers, soldiers and civilians.[62] The multiple consequences were found across the field of conflict.

As from the outset, money was a key element and involved drawing on the income and credit possibilities of a range of wealth-holders, both secular and ecclesiastical. Thus, the Crusades had seen many mortgages on landed estates in order to raise money. Moreover, across the world, albeit in different forms, merchants were able to draw on far-flung credit networks. Fourteenth-century military contractors had benefited from the use by Aragon and Florence of financial incentives in order to organise logistical support, and evidence of such systems increased.[63] In large part, this was an aspect of a longer-term pattern of incentivising support by means of money rather than other forms of benefit, notably pillage and land. However, that Parliament introduced impressive fiscal devices, especially the assessment and excises, no more made its victory certain in the English Civil War or that of the Cromwellian regime against Spain in 1654–60, than the efforts of other powers to increase their fiscal yield and control.[64]

At the same time, as also for example with the case of fortifications, the shifting boundaries of 'public' and 'private' are at issue here, and, in addition, the dangers of applying modern categories of identity, obligation and consent on the past. These points ensure that the location and evaluation of logistics is in part a matter of broader questions of political thought, and these were highly specific to particular territories. Those readily applying modern notions of rent-seeking and public-choice economics to the past need to address these contextual historical points before the notions can be applied successfully. That issue is an aspect of the more general problem with modernisation theory.

Politically, for example, France, a very different polity to that of city states, has been widely treated as a success and, under the administrations successively of Cardinals Richelieu and Mazarin (1624–61), as 'the early pattern of the state-controlled and administered army which was to triumph under Louis XIV [r. 1643–1715]'.[65] However, such an army was arguably less effective logistically than those reliant on contracted relationships with private entrepreneurs,[66] relationships which permitted a drawing on wide-ranging networks of credit and supplies. These ensured that the pressure on local areas was relaxed by being spread, and this enabled a continuation of effective operational warfare.[67] The French Crown, in contrast, for reasons of political control and image in the aftermath of the Wars of Religion (1562–98) and of later aristocratic conspiracies, sought to maintain direct royal control over the army and its supplies, only for its method to prove largely ineffective[68] and heavily reliant on contributions.[69] Logistical failures in the French army led to massive desertion, a clear index of failure, but one also seen with other armies whatever their method of support.

More mundanely, continuing the changes earlier seen with wagons, benefit came from the spread in the sixteenth century of mobile front-wheel axle units for the four-wheeled wagons that predominated in northern Europe. There were also improvements in ship construction and specifications, again with considerable overlap across the chronological divides generally applied in history, notably, for Europe, Antiquity, the Middle Ages, and the Early-Modern period. Moreover, the course of campaigning underlines the extent to which the period of the supposed European military revolution did not see any logistical revolution; none for example in the construction or locomotion of vehicles, nor in road surfaces and bridging techniques. Indeed, in eighteenth-century Tuscany, a wagon drawn by four horses pulling 4,000 lb could rarely cover more than 20 miles daily. Poorly-constructed roads often enforced the use of light carts and only two horses, increasing the number of carts necessary to move a given load and the consequent cost in manpower and forage. Still more often, burdens were limited to 2.5 cwt (280 lb) or so, which could be carried in panniers on a horse or mule, against the 10 cwt (1,120 lbs) which could be drawn by a single horse over good roads.[70]

Road improvement could therefore greatly improve loads, but often wagons and carts afforded merchandise only inadequate shelter, while loading, packing and unloading techniques were simple. Moreover, there could be particular problems, as in 1748 when Venetian contractors were unable to find the 5,000 mules required by the Austrian army in Italy then campaigning against French, Spanish and Genoese forces, or in 1708 when a rainy summer made the Russian and Lithuanian roads very soft, hindering the Swedish invasion of Russia. The Swedes did not have the possibility to use a river route, as they had done in invading Poland with the Vistula in 1703 when moving supplies and artillery. More generally, any improvement in logistical capability was outweighed by persistent resource problems. From Antiquity to the mid-eighteenth century, the theme is of small incremental improvements in underlying technology and economics, but, crucially, with the fundamental constraints basically unchanged.

The limited effectiveness of states understood as central governments, and their reliance on mixed systems of support, were seen in the aftermath of wars as well as during them as it was difficult to meet arrears in military pay, as well as to demobilise forces, and therefore there was a tendency to search for expedients. Thus, after the Thirty Years' War ended in 1648, Sweden maintained garrisons in Germany until 1654 to secure payments, and then sought to move the burden of this support by invading Poland; although far more than simply this factor was involved in the invasion.

Money, the key means of logistical capability, was largely available only so as to permit the operation of one effective campaign army, and this tended to be the case whatever the supply system. Thus, the Spanish focus in 1629–30 on the War of the Mantuan Succession in northern Italy left the Spanish Army of Flanders short of funds, and therefore with its operational effectiveness compromised and the army downright mutinous. In 1658, Tsar Alexis of Russia signed a truce with the Swedes so that he could focus on the Poles. In 1635, the French armies that invaded the Spanish Netherlands and the Valtelline valley of northern Italy were weakened by serious supply failures. Henri (II), Duke of Rohan, who commanded the latter, discovered that his troops were no better supplied than when he had taken part in the Huguenot rebellion of 1625–9, and, short of supplies, his army collapsed in 1636–7.

Mobilisation and logistics relied on a decentralisation of authority and a role for market forces. The creation and supply of large armies and fleets using partnership arrangements with military entrepreneurs, many of them aristocratic officers, as well as with commercial and financial interests, showed the originality in thinking and organisational flexibility of the period, as did the use of state-condoned predation on land and sea through contributions and privateering. At the same time, the military and political costs of conflict were such that resources and commitments had to be carefully balanced, with logistics a dependent part of this difficult and inherently subjective equation.[71] Moreover, the often limited control that the Crown had over armies on campaign helped ensure that logistics in part was a matter of maintaining the cohesion both of the armies and of their links with the Crown. This dimension was more generally true of the politics of logistics, and notably so with the use of pay. Again the 'total' nature of logistics as a subject emerges.

Chapter 3

From the End of Ming China to the Fall of Safavid Persia, 1644–1722

The period from 1644 to 1722 deliberately selects two dates that do not feature in European military history, but the fall of two long-standing empires, Ming China and Safavid Persia, represented both significant moments and also major triumphs for opponents who did not rely on large forces of infantry supplied by bureaucratic means. To adopt this periodisation therefore calls in question the standard analysis and narrative of logistics. For this period, the latter would focus for example on the development of magazines or stores, particularly by the French in the late seventeenth century, and by John, Duke of Marlborough when, during the War of the Spanish Succession (1701–14; 1702–13 for Britain), successfully moving his troops from the Low Counties to the Danube in the Blenheim campaign of 1704. These achievements were indeed significant, but also need to be set in the context of the situation outside Europe.

France, the most populous European state other than Russia, developed a logistical system with the supply networks of men, money, munitions and provisions improved as part of what has been referred to as a state commission army rather than an aggregate contract one.[1] Alongside a supporting system of *étapes* (supply depots along marching routes), there was a network of magazines near France's frontiers from which campaigns could be supplied. This network was used with considerable success in launching the War of Devolution and the Dutch War in 1667 and 1672 respectively. Thanks to these magazines, the French could seize the initiative by beginning campaigns early. These achievements owed much to two successive talented Secretaries of War, Michel Le Tellier and his son, François, Marquis de Louvois, who, in succession, held office from 1643 to 1691. They presided over a relatively well-organised War Office; although one whose reputation has been lessened by recent scholarship which has indicated for example serious corruption. In turn, these magazines had to be fortified, and thus gave a purpose to the French fortress system that is particularly associated with Sébastien Le Prestre de Vauban, for example the major fortress-magazine at Lille.[2]

And so also with Marlborough's campaigns, which reflected impressive organisation alongside the financial strength of Britain and the Dutch: in 1704, his army advanced 350 miles from Bedburg, between Ruremonde and Cologne, via Mainz to Launsheim, the most decisive British military move on the Continent until the twentieth century. Depots of supplies were established along the route, providing the troops with fresh boots as well as food, and enabling the army to maintain cohesion and discipline, instead of having to disperse to obtain supplies which would also have been politically unwelcome because the advance was largely through allied territory. The presence of supplies en route reflected excellent organisation and also the gold supplied to the contractors from Britain, gold a product of Britain's global trading system, and notably of obtaining gold from the Portuguese colony of Brazil. Marlborough's subsequent move back to the Low Countries, however, in part reflected the greater ease of obtaining supplies there, both from Britain and locally, although the politics and requirements of strategy were more significant. The bread contracts for Marlborough's army yielded a percentage to Marlborough. He also benefited from a first-rate Quartermaster-General, William Cadogan.

That so much campaigning was concentrated in particular areas, for example the Low Countries and Lombardy, encouraged an analysis that emphasised the congruence of troop numbers, food capacity and administrative capability, with the supply of bread crucial for operational reasons.[3] So also with naval supplies, where there was local procurement, but also complex long-range networks, notably for wood (of certain specifications), hemp, tar and iron. These networks linked Baltic sources of naval supplies to shipyards in Western Europe, as well as Balkan timber supplies to Mediterranean shipyards, notably in Venice.

Only certain states were able to draw on the wideranging supply networks of maritime stores seen as important to English naval power by Thomas Hale in 1691,[4] networks that helped explain the strategic importance of particular areas for construction and maintenance and that were an aspect of the wartime struggle over logistics. Food supplies were also of strategic importance, as well as operational and tactical, and thus their interruption and safeguarding were significant, as in 1694 in the North Sea when a fleet of grain ships, *en route* from the Baltic to France after two years of poor harvests there, was captured by the Dutch, only to be regained by the French. This was not the sole source of grain transported by sea. In the 1690s, Spain continued its practice of providing grain from its possessions in Naples and Sicily to the contractors preparing *pan de municion* for its army in the Milanese.[5]

The model of effectiveness offered by the standard focus on the French and the English/British provided a way to assess other states, and also, linked to this, part of a developmental model through time. In addition, the discussion of logistics in terms of organised supply trains maintained from magazines provided a significant basis for an assessment of operational factors. In particular, bridging points were key nexuses for systems reliant both on crossing rivers and on moving large quantities of supplies along them, such as the Meuse in the Low Countries.[6]

This account deserves consideration, but care is required before adopting it as a prototype for progress, or indeed a basis for judging developments across the world. Two particular arguments will be offered here. First, the standard account of developments in Europe rests, as more generally with the assessment of power and government in the so-called 'Age of Absolutism', on an analysis that does not sufficiently qualify the capability of the state. In particular, there is a widespread failure, notably by non-specialists, to understand adequately the extent to which rule and government in the 'Age of Absolutism', essentially self-government (other than in religion) at the king's command, rested on consensus and co-operation with the élite. The latter often played a key role, not only as officers but also as the source of loans. In many cases, officers from the social élite played a central role in recruiting, feeding, housing and transporting forces, either on their own responsibility or through their role in local government.[7] Whatever the situation, it is necessary when discussing the bureaucracy of the 'Age of Absolutism' to draw attention to what should be anticipated, namely frequent supply issues, issues that need to be considered because they undercut the standard view.

Secondly, the political dimension of campaigns helped ensure that general judgments about capability have to be placed in specific contexts, notably those of particular political goals and advantages. This was seen in 1690 when, having seized power in England in 1688–9 as part of the

'Glorious Revolution', William III advanced on James II in Ireland. The French advice to their ally James was to play a waiting game, one of war by logistics, rather than to risk battle. James was urged to burn Dublin, to destroy all the food and forage in William's path, and to wait for a French fleet to interrupt William's seaborne supply route from England, and for subsequent privation to demoralise and weaken him, but James understood the need to consider political as well as military contexts. Grasping Dublin's symbolic and strategic significance, and fearing that the political and logistical strains of delay would lead the Irish and French forces to waste away before his opponents, James decided to fight, only to be outmanoeuvred and totally defeated by William at the Battle of the Boyne.

More generally, developments in warfare in Europe were mediated through existing social and political structures, rather than being imposed on them by 'the state'. Thus, the nobility remained the leading military group,[8] while the resources of government were affected by these structures, both positively and negatively. An alternative to contributions was that of state finances, by taxation or loans, as used by the Dutch Republic in a representative political form that included a measure of consent.[9]

Civil war, with its potential for significant change, a basis of the creation of the Dutch Republic in the late sixteenth century, could lead to a new social politics that made possible a different military, as with the New Model Army created in England by the Parliamentarians in 1644. This well-disciplined force, which served as the expression of the political thrust of the English revolution as well as providing its physical power, was supported by a more effective infrastructure and supply system.[10] When, under Oliver Cromwell, who, like Marlborough and Wellington was impressive at logistics, the New Model invaded Ireland in 1649, it was well-supplied, and thus able to operate most of the year round; well-equipped with carts, wagons and draft horses, the army retained the initiative.[11] This, however, was an unusual type of army, one that took over the country's government, but could not be politically sustained as it proved impossible to maintain the political situation created by Cromwell. So also with many other instances across history in which a revolutionary army was to make a new state, but be unable to sustain it.

To make a comparison with a very different force, the Afghan Ghilji, who completely overthrew the Safavid empire in 1722, were also a highly effective military, but, again, they were unable to sustain the new system they created in Persia, and their rule was overthrown in 1729. The contrasts between the governmental systems in England and Persia under the new regimes are readily apparent, but there was also a similarity in the shape of new dynasties: the Hotaks (Ghilji rulers) in Persia from 1722 to 1729, and the House of Cromwell, in the persons of Oliver and his son Richard, as Lord Protector from 1653 to 1659. Dynasticism may seem a redundant form of governance, but it can be seen in North Korea and Syria; while monarchy of a form is present with longlasting dictatorships as in Cameroon.

In England, the New Model Army proved less effective in sustaining the new system than the military force that was to be created under the Revolution Settlement that followed the 'Glorious Revolution' of 1688–9, a new political system following the successful Dutch invasion of 1688, but a system clothed with the trappings of continuity. At the same time, there were continuities, notably with the navy: from expansion under the Interregnum (1649–60) to development under the Restoration Monarchy (1660–88), for example with the establishment of the Victualling Board, and then to fresh expansion in English (from 1707 British) logistical capability after 1688–9.

That internal stability was significant to external power projection not only in ensuring sufficient resources on a longterm basis, but also in enabling the dispatch of troops, was to be shown in China after the end of the Rebellion of the Three Feudatories (1673–81). This permitted the determined Kangxi Emperor (r. 1662–1723) to turn against the Zunghars, the latest (and last) of the major steppe challenges to China.

So also in India after the warfare over the Mughal imperial succession in 1658–9. The victor, Aurangzeb (r. 1658–1707), who during his reign was often at war, controlled a formidable military system that had an impressive logistical capability. This capability was enhanced by roads, for example the improved and renamed 'Imperial Road' from Lahore to Kashmir that was the product of the conquest of Kashmir in 1589 by an earlier Mughal Emperor, Akbar (r. 1556–1605). Akbar also improved the road through the Khyber Pass. New roads under Aurangzeb an all-weather road from Shangramgarh to Dacca in order to support operations on the Bay of Bengal. These logistics had to support operations in a variety of environments, from the equatorial heat and humidity of the forested Brahmaputra valley, where the Mughals fought the Ahom, to the arid plains and snowy mountains of Central Asia.

This range of activity provides a measure of logistical capability that was similar to that of Ancient Rome or the British army in the early nineteenth century and different to those of the many Western forces, that largely operated only in one type of military environment, such as those of Frederick the Great of Prussia (r. 1740–86) and Moltke the Elder, the key general in the German Wars of Unification (1864–71). The Mughals had to follow different policies in the physical and military environments to which they needed to respond. Mughal armies faced problems in transporting siege artillery to the Deccan, notably the lack of navigable rivers. In Assam, in contrast, the Mughals were able to use the Brahmaputra River.

At the same time, the Mughals faced the problems posed by asymmetrical warfare, which included that of asymmetrical logistics. The Marathas, a Hindu warrior caste in the western Deccan around Pune (Poona), became a major opponent in the late seventeenth century. Just as their tactics relied on avoiding the shock power of the heavily-armoured Mughal cavalry, instead weakening them by the use of missiles, so the Marathas' operational preference was one for gaining logistical mastery by cutting the supply lines of opponents, and a related strategy of exhausting them by devastating territory rather than risking battle. Supply lines should be understood as routes along which supplies were moved, generally in wagons or carts.

Similarly, the Poles relied, as in successive wars in 1626–9 and 1655–6, on attacks on Swedish supply lines in order to impede their operations, thus forcing the Swedes to peace in 1629 and to withdrawal from Poland in 1656. Anger at Swedish 'contributions' had led to a massive increase in resistance in Poland from the autumn of 1655.[12] Such tactics encouraged a stalemate in which both sides could launch attacks, but not hold the ground the attacks gained.

Exhaustion was a problem for armies in this context, as with the Russians against Poland in the early 1660s and the Austrian army in the Rhineland in 1677. William Skelton, the English envoy in Vienna, noted of the latter. In May, that its commander 'wants ammunition and cannon … besides he will ruin his army if he goes from the Rhine but four days march', adding in September 'where money and bread is wanting no general can do wonders'.[13] Troops were dispersed to forage and to man positions protecting supply zones and many, mostly small-scale, encounters arose from raids and from related (but also separate) attempts to define contribution zones.

This situation helped put a premium on the value of 'small warfare' in order to serve operational goals, with raids taking on meaning within the calculus of supplies and also as a means of putting pressure on hostile garrisons. Thus, in July 1678, a force of about 1,000 Dutch troops sent from the fortress of Mons to escort in a supply convoy beat a similar French force that tried to stop it. Indeed, that aspect of logistical warfare was central to the offensive-defensive character of so much war making. A classic instance was on the French border with the Spanish Netherlands (Belgium) during the conflict of 1673–8, and this instance, brought to the fore by recent scholarship that raises the question of what other such work could reveal, stands alongside the more usual scholarly emphasis on large-scale battle and siege. The French tried to protect their own northern frontier, as well as to put pressure on the Spaniards. It was impossible to seal off the frontier, but the French authorities dispatched war parties, prepared defences, issued ordinances, cajoled local officials, and organised militia; all with the intention of slowing and disrupting the Spanish imposition and collection of contributions. Moreover, French reprisal raids and escalating demands for contributions were intended to inflict harm on the inhabitants of the Spanish Netherlands, and to weaken their resolve. The French defences worked well, but, in a form of pro-active logistics, driving the Spanish garrisons from their fortresses was the only sure means to prevent Spanish raids. To that end, blockades of fortresses were an important prelude to sieges and demonstrated the potency of French 'small war'. These blockades, which, crucially, continued during the period of winter quarters in what was, in this respect, an attritional war of outposts, wore down the Spaniards, enervating their garrisons, or ensuring that French sieges were more rapid.[14]

Yet, logistics was on a wider spectrum, with sieges in that region also affected by supplies that had to be brought a considerable distance, and therefore on the factors involved, including the impact of a lack of water on river levels and, thus, on the ability to move cannon by that means.[15]

Alongside such functional deficiencies can be found the consequences of a social and cultural politics in which the state was understood widely in dynastic and élite terms[16] that might be termed traditional were it not that similar features can be found in many states today. Political culture and functional factors interacted, with consequences noted in an English memoir on French resources, written in 1678 after a tour of the country, suggesting that, although the army had 265,000 troops listed as in pay, at least a fifth 'may be deducted by false musters and other devices of officers, notwithstanding all the great rigour used against those that are found faulty'.[17]

Given that all systems faced deficiencies, a situation that is still very apparent today and that will continue to be so into the future (on which see chapter 11), it was, as ever, the relative situation that was to the fore. This was seen in the frequent conflicts between Austria and the Ottomans where logistical effectiveness was important to operational capability in part due to the lack of local supplies in the extensive devastated war zone. The Austrian system, based on the *Proviandhausen*, depots scattered along the marching routes, helped in the conquest of Hungary,[18] but was less effective than the Ottoman one, although Ottoman preparations for the siege of Vienna in 1683 were inadequate.[19] Each of the systems, coming down or up the Danube, was supplied by water but, whereas the Ottomans had an effective river flotilla, the Austrians only developed a good one around 1715. Moreover, after 1683, the Ottomans were fighting in their own territories on the defensive, and therefore closer to the ordnance and supplies they had previously accumulated. The first Austrian campaign in the Balkans that was logistically well-organised was that of 1716.

Prior to that, every Austrian campaign, including all those fought in Italy during the War of the Spanish Succession (1701–14), was logistically unsatisfactory, with the troops almost starving and short of shoes and horses. When campaigning in Spain, there was a dependence on British financing, notably the resources created by long-term, lower interest borrowing.[20] The Austrians prevailed in Hungary in 1684–99 not due to better logistics – indeed they were just bearable, unlike say those of the Russian forces that unsuccessfully attacked the Crimean Tatars in 1687 and 1689. Instead, the Austrian commanders, notably Duke Charles V of Lorraine, Louis William, Margrave of Baden-Baden, and Prince

Eugene, were on the whole superior to those of the Ottomans, while they were also part of an alliance including Russia, Poland and Venice, that divided the poorly-commanded Ottoman response.

The same period saw China deploy a formidable army that arose from the fusion of the earlier Ming system of a classic sedentary, big state, infantry army with the added capability and determination derived from Manchu takeover and the steppe cavalry forces thereby added. This new system, which incorporated different logistical requirements, proved capable of operating beyond the bounds of its Ming predecessor. In 1683–9, Chinese forces drove the Russians from both the Amur basin and from lands to the north, sustaining sieges of the Russian fortress at Albazin. Then the Chinese were brought into confrontation with the Zunghar Confederation of west Mongolian tribes, a more formidable adversary. In 1690, the Chinese advanced, fighting the Zunghars at Ulan Butong, 320 kilometres north of Beijing, but a shortage of food made the Chinese commander happy to negotiate a truce.

In 1696, the Kangxi emperor, a determined figure, advanced north across the Gobi Desert with converging columns, a serious test of logistical capability and 'near-run thing' which led his advisers to urge him to turn back before the army starved. The Chinese, however, coped far better than the Russians had in their advance across the steppes against the Crimean Tatars in 1689, and the Zunhgar army was destroyed in battle at Jao Modo. Underlining the need for a multi-faceted explanation, success owed much to support for the Chinese from the rebellious nephew of the Zunghar leader. Yet, Chinese effectiveness also reflected the strength of the army including an underlying impressive logistical system ensuring that the wealthy agricultural zone of lowland eastern China could be tapped to support campaigns to the north and west closer to the frontier. The combination, under the Manchu, of steppe and Chinese forces and systems, had proved crucial in ensuring that the frontier was overcome, a process underlined when settlement was undertaken in conquered areas in order to provide resources to support the army.[21]

The Chinese could fail logistically, as with unsuccessful advances on Zunghar-held Lhasa, the capital of Tibet, in 1717 and 1718 that ran out of food, but in 1720 they were more successful and captured Lhasa. Another formidable display of power projection occurred to the north of Tibet where the Chinese and the Zunghars contested the oasis settlements that were crucial bases on the trade route to the west as well as vital supply-points. The Chinese captured Barkol in 1716, Turfan in 1720, and Urumchi in 1722. But, again, logistics was only an enabler. Thus, Chinese operations were helped by a rebellion by the Muslims of Turfan against Zunghar control.[22]

Although as already indicated, the Mughals displayed a multiple effectiveness in campaigning in different environments, they could not match Chinese success. The conquest of the Deccan was not matched in Kandahar or northern Afghanistan. Moreover, their Maratha opponents avoided battle, preferring their strategy of *bargi-giri*: cutting supply links, launching devastating raids, and using hit-and-run tactics. However, as with the Kangxi emperor, logistical capability was significant. Thus, the Mughal presence in Bengal was strengthened in the 1690s by the construction of an embanked road to Alamgirnagar (Cooch Behar). Against the Marathas, many of the troops and animals, and much of the supplies and money for Mughal operations, were obtained in the north and then moved south on the royal road from Delhi to Burhanpur. Unlike Kangxi against the Zunghars, it was possible for Aurangzeb in his operations to draw on local and nearby agricultural resources, but he confronted the problem of large-scale and lengthy sieges. Whatever the deficiencies of the Mughal regime in Aurangzeb's last years, European rulers would have considered themselves very fortunate to wield such power, as the impressive range of Mughal sway rested on an effective military infrastructure. So also with the Ottomans, as in the rapid and successful conquest of the Venetian-ruled Morea (Peloponnese) in 1715.

Again directing attention away from Western forces, relative capability was seen in 1711, when Peter the Great invaded Moldavia. The Russians, who in Turkish eyes had not provisioned the expedition as well as the unsuccessful Russian ones against Crimea in 1687 and 1689,[23] were affected by the need to operate at a considerable distance. By advancing as a single force, which provided a margin of safety, the Russians increased their already serious logistical problems, which were exacerbated by the failure of the Moldavian harvest, and, once surrounded, short of food and water, Peter had to accept humiliating terms.

The significance of supplies increased that of sieges which, in turn, underlined the importance of logistics, providing a focus for activity rather like, albeit in a different fashion, as the front lines of the First World

War were to do. Fortresses both secured territorial gains and dominated supply routes. Belgrade and Temesvár controlled both the land and river routes necessary to deal with the shortage of available supplies in Hungary. The Austrians used cannon in large numbers from the second half of the seventeenth century, and, in 1716, their commander in Hungary, Prince Eugene, had 90 field cannon and about 100 siege guns. Artillery helped make armies slow-moving, which enhanced the importance of supplies. In addition, for the Russians on the steppe to the north of the Black Sea, the limited amount of food available underlined the significance of logistics. Large supply trains were necessary there, as was the use of river routes to move food. Thus, sieges undermined the prospect of moving swiftly, instead forcing a different, more methodical, taking of the initiative from opponents. Whichever method was pursued by combatants certainly depended on a sound grasp of logistics, but the latter no more dictated the method of the Crimean Tatars than that of the Russian army.

Moreover, the emphasis on the regular forces of developed states plus their logistics looks somewhat confusing given the advance of the Ghilijis of western Afghanistan. Rebels against the Safavid empire, they captured the cities of Kandahar (1709) and Kirman (1719), before advancing into central Persia in 1722, defeating the Persians at Gulnabad and then besieging and capturing nearby Isfahan, the capital. The empire collapsed, as it had never done when attacked by the Ottomans. That might appear to signal an opportunity for Western intervention, but, in 1722–3, when Peter the Great of Russia advanced along the Caspian Sea, the logistics, despite an officer having examined the roads in advance, were poorly-organised and inadequate, notably the supply of food and ammunition.

So also, in June 1711 with the report from John, 2nd Duke of Argyll, the British commander of an allied army in Barcelona during the War of Spanish Succession. With his Spanish troops mutinying for lack of pay and his own forces short of cannon and powder,

> having with greater difficulty than can be expressed found credit to keep the troops from starving in their quarters all this while, which for my part I do not see how we shall be able to do any longer, for the not paying the bills that were drawn from hence the last year, has entirely destroyed her Majesty's [Queen Anne's] credit in this

place; but though the troops could be supplied in quarters, that will not now do the business, for the enemy is already in motion ... so that if we remain in quarters, we shall be destroyed *en detaille*, and to get together is not in nature till we have money, for the whole body of troops that were here last year are without all manner of necessarys, having both officers and soldiers lost all their tents, baggage and equipage at the battle of Villaviciosa [10 December 1710], besides that the contractors for the mules to draw the artillery and ammunition and carry the bread will by no means be persuaded to serve any more till we have money to pay them.[24]

Logistics was far more serious in Spain than in the Low Countries, not only because of the relative poverty and shortage of food in Spain and Portugal,[25] factors which also affected Napoleon, but also because the British ally in the Low Countries (the Dutch) was far better able to supply its own forces and to pay a portion of the cost of financing allied troops. It would be misleading to use Argyll's letter to characterise the British situation as a whole, but, equally, it should be borne in mind when the example of Marlborough's march to the Danube is cited alone, as is usually the case. So also with the difficulties of assessing the effectiveness of the increasingly-strained French army administration.[26]

Royal control in part depended on the willingness of the officer corps to accept subordination while also bearing part of the cost, a situation that continued during the *ancien régime*. The emphasis on Absolutism as the basis for a new logistical framework might profit from considering why previous emperors and kings who derived authority from God still faced great resistance from the aristocracy. So also with that from areas distant from the centre of power. Thus, what legitimacy and authority entailed were matters of opinion and practice. While socio-political collaboration was a key element of the costly state monopoly of force,[27] so differently, were contributions, both from foreign populations and from that of France, the latter largely in the form of payments by provincial treasuries for quartering troops. The billeting of soldiers on the populace was a form of 'tax of violence' that was unpopular and, therefore, came with political costs. It could also be employed as a form of control, as with billeting on Huguenots (French Protestants) in order to intimidate them into conversion to Catholicism.

More generally, the key element of long-distance combat, logistics, remained poorly organised and, in part, arbitrary in its operation and unpredictable in its success. As a product in part of 'logistical geography',[28] the presence of the king, for example Louis XIII and Louis XIV, with a particular army was usually highly significant in keeping it reasonably funded and supplied, and thus more likely to succeed. The king also helped settle difficulties between officers over precedence. In many respects, this was not so much a 'modern' characteristic as a continuance of a long-established pre-modern one. Military success brought *gloire* to the monarch and also helped him employ, encourage, control, and discipline the social élite. On the long-standing pattern of monarchy, seen for example of the Chola (south Indian) kings Raja Raja and Rajendra in the eleventh century, this was true of Louis XIV, Peter the Great, and the Kangxi Emperor, and was also to be true of Napoleon, whose military rule was deeply traditional in this respect, and in many others.

Discussion of this period in terms of large armies and their logistical requirements is in line with the general tendency of the scholarship, but, that tendency has its disadvantages in assuming a normal and/or desirable form of army and military activity which, in practice, neglects the range of both. The equivalent to van Creveld's work for this period is the Géza Perjés article in footnote 3, an oft-cited work that suffers from the drawbacks of its assumptions and that should not be extrapolated as a formulaic model without understanding the significance of particular environments, in terms of both constraints and responses. The varied ability to respond to these specific factors emerges clearly from the warmaking of particular empires, for example those of the Mughals and Manchu.

Chapter 4

From the Fall of Safavid Persia to the Outbreak of the French Revolutionary Wars, 1722–92

'As to the European powers their want of money is the best security for the preservation of the Peace ... they cannot make war without nerves to support it.'[1]

Joseph Yorke, British envoy in The Hague, 1763.

The most impressive large-scale logistical achievements in the period were those of the Chinese army in Central Asia and of the British navy, the Royal Navy, each formidable organisational systems that were products of institutional continuity, political commitment, and economic strength. The crucial factor in Chinese military capability was not weaponry, but the ability to deliver and sustain considerable power at a great distance having made preparations. This situation matched that within the Western world where organisational developments, range and capability were more important than military technology and tactics (which Western powers shared, notably at sea), and in terms of both absolute and relative power.

Yet again, however, the logistical achievement, while necessary, was not in itself sufficient for the Chinese, although it was important to the management of steppe warfare. In the 1750s, building on their earlier system, the Chinese established two chains of magazine posts along the main roads on which they advanced in Central Asia against the Zunghars. Supplies were transported for thousands of miles, and the Mongolian homelands controlled by their eastern Mongol allies provided the horses and plentiful fodder, each crucial to operating on the steppe.[2] These improvements in logistics – due partly to a desire to keep the troops from alienating the populace and partly to the latter's very lack of food in a very arid region – ensured that the Chinese armies did not disintegrate as Napoleon's did in 1812 when he encountered serious problems in invading Russia despite advancing over a shorter distance. Comparisons are difficult, not least because Napoleon faced greater resistance than the Manchu armies had done in 1755, but the contrast in the supply situation was very important. It is clearly inappropriate to put the Europeans to the fore in logistical discussion.

The application to military purposes of the great demographic and agricultural expansion of China during the century was related to the extension of arable farming in Gansu, and, in order to ground the new conquests with the establishment of colonists, shifting the focus in these areas from animal husbandry.[3] In order to wage war with the Zunghars, there was a massive transfer of resources from eastern to western China, and, as with other instances of Chinese warmaking, this capacity reflected both the administrative capability of the state and the extensive resources of well-developed mercantile networks. The Manchu conquest of Central Asia succeeded in part because they successfully solved the logistical problems which no previous dynasty could surmount, although other factors were involved, notably civil war and defections among the Zunghars and also devastating smallpox. The combined Manchu, Mongol, and Han, infantry, artillery, and cavalry forces, discovered how to manage steppe warfare, because that was considered the central strategic threat by all Chinese dynasties, while they were much less successful on the Burma, Sichuan, Vietnam, southern frontiers both because those were not of central strategic interest and because the physiographic environment was quite different.[4]

As a reminder of the difficulties of judging effectiveness, alongside the capability of the Chinese government system, it was moreover affected by limitations, including the leakage of tax revenue to officials and the extent of tax farming.[5] But this situation also helped in drawing on non-governmental resources, and, for both financial and organisational reasons, the extent of commercialisation and market integration in the Chinese economy was important. Economic strength and logistical capability had foci and links in the stability and range of entrepreneurial networks, and, related, rather than separate, to that, in the relative effectiveness of state finances,[6] and to a greater degree than with the other major Asian states.

Chinese methods alone did not bring success, as was amply demonstrated with repeated failures in attacks on Burma in 1765–70, and later against

Vietnam in 1788–9. Again, a range of factors were at play, notably the many consequences of the unfamiliar, heavily-forested, environment that was difficult for large-scale military operations, and also rife with diseases, particularly malaria, affecting troops and horses. The problems posed for logistics, therefore, were but part of a wider difficulty. At the same time, Chinese flexibility was shown in 1769 in the building of hundreds of boats in Yunnan in order to help in the unsuccessful invasion.[7]

These results, very different to that against the Zunghars, underline the problem of any reading from logistical capability, both absolute and relative, to success – or failure. Similarly, failures against Burma and Vietnam did not mean the same against Nepal from which Gurkha attacks on Tibet had been mounted in 1788 and 1790. In 1792, the Chinese dispatched troops, mostly from nearby Sichuan and Tibet, but including bannermen (élite units) from more distant Beijing and Mongolia. This was a formidable and costly logistical undertaking that owed much to financial support from the major salt merchants who provided much of the money for military operations. The expedition suffered heavy casualties and was challenged logistically and by the approaching winter; but the Chinese advanced to near Katmandu and the Gurkhas offered acceptable terms including a tributary relationship.

In one respect, this is an appropriate place to end the chapter before switching to the outbreak of the French Revolutionary Wars in 1792. Such a divide captures the marginal significance of land conflict in Europe, a point that can be taken further by emphasising, instead, that the European powers that were of importance on the world scale were Russia, because of its operations against non-Western powers, notably success against the Ottomans and presence to the Pacific, and the trans-oceanic Western imperial states. Any comments on say Frederick II, the Great, of Prussia (r. 1740–86) could then be relegated to relative obscurity as of only regional significance, alongside, for example, those of the conflicts of Burma and Siam (Thailand). This point should indeed be brought to the fore, because it is all-too-easy to focus instead on Europe as in van Creveld's work. Frederick II's 'just enough' logistics were the basis of a later German practice that led in a different economic and political context to German disaster in both world wars. Essentially prepared only for short wars, his armies were launched into wealthier agrarian regions, notably Silesia, Saxony, and Bohemia, obtaining supplies from them. However, when this course was blocked, particularly in the War of the Bavarian Succession (1778–9), then poor logistics helped wreck the Prussian campaign.

In contrast to the limited range of Frederick II and of other rulers of essentially regional significance, Western naval power rested on a public finances[8] and an infrastructure that enabled specialist warships to operate around the world, with the British, who had the world's leading navy from the 1690s to the 1940s, indeed capturing Havana and Manila in 1762. Naval bases were founded and existing ones enhanced, so that a new geography of naval power, based on ports such as Plymouth and Brest, was created. Both these ports had direct access to the Atlantic, which became more important to Britain and France than had been locations on the North Sea and the Mediterranean, such as Chatham and Toulon respectively. Dockyards were among the largest industrial plants, employers of labour, and groups of buildings in the world, and were supported by massive storehouses, such as the vast Lands Zeemagazijn in Amsterdam. Even minor naval powers developed an infrastructure, Savoy-Piedmont establishing yards and other buildings at the naval headquarters, Villafranca, as well as at Nice and on the island of Sardinia.

British capability rested more generally on the strength of public finances and on a greater commitment of national resources to naval rather than land warfare, a political choice that reflected the major role of trade, the character of the national self-image, and geography, with Britain an island no longer under land threat from Scotland.[9] British naval efficiency was to increase from the 1790s with the introduction of particular attitudes and practices of financial and administrative control, but this introduction reflected what was already present: a more general openness to new thinking and a readiness to consult those with expert knowledge.[10]

As a central aspect of developing the oceans as a 'manoeuvre space', as well as indicating the role of bases, the Europeans also took their naval-military industrial capability outside Europe, with major shipyards at colonial bases, and indeed bases such as Havana, Kingston, and Bombay (Mumbai) as fiscal-military hubs. The growing British naval and mercantile presence in the Indian Ocean owed much to shipyards in India where ships could be built and repaired, especially Bombay where, by the mid-1770s, the dry dock, the sole one available to the Royal Navy outside Britain, could take three third-rate warships.[11] Moreover, the Navy Board, the Victualling Board, and the Ordnance Board, were able effectively to direct a formidable re-supply system, including on the far-flung East Indies Station, as well as resupplying at sea the warships blockading the crucial French Atlantic naval base of Bristol. This close blockade was the necessary background to the destruction in 1759 of that fleet at the battle of Quiberon Bay, which ended French invasion hopes during the Seven Years' War.[12] Naval strength was also the basis for power projection on land, and not only in India. At the same time, naval strength and maritime capability could only do so much. The British conquest of Canada in 1758–60 required a formidable logistical underpinning on land as well.[13]

A very different pattern of behaviour was that of Persian and Afghan warfare. Offensive operations, those for example of Nader Shah of Persia (r. 1736–47) and Ahmad Khan Abdali, the founder of the Durrani empire based in Afghanistan (r. 1747–73), were in part designed to gain spoils in order to reward troops and to transfer the burden of support, a practice also seen with the Turks ravaging Habsburg Wallachia and Serbia in early 1738. Moreover, on the part of Nader and the Afghans, there was a preference for battles rather than sieges because of the logistical burden of the latter. This might appear a primitive model of achievement, not least because the basis of state authority was weak. However, by 1740, Nader (a Turcoman tribesman who was the key general from 1729) had expanded Persian power further than any of the Safavids, even Shah Abbas, defeating the Turks in battle more consistently than they ever managed, gaining an entry into India (capturing Delhi in 1739) the Safavids had not enjoyed (nor any Europeans at this stage), and subjugating Central Asian cities like Khiva, Bukhara and Samarkand that the Safavids had never reached. By 1743, he had a powerful and well-supplied new artillery train.[14]

Nader's campaigns demonstrated the value of cavalry, as did the Marathas who proved able to cut off the grain supplies and reinforcements of rival Mughal armies, as in 1728 and 1735, and also those of the army of the Nizam of Hyderabad. Similarly, in 1769, the light cavalry of Haidar Ali of Mysore ravaged the Carnatic. This was a pattern also seen elsewhere. The cavalry of the West African state of Oyo pillaged Dahomey between 1726 and 1748, forcing it to pay tribute.

In 1735, the surrounded Mughal Emperor, Muhammad Shah, was forced to buy the Marathas off with a cash tribute, which represented another way to raise logistical support. So also in 1739 to buy off war with Nadir Shah, while Muhammad Shah's successor, Ahmad (r. 1748–54), bought off the Afghans in 1750; and, in 1751, Alivardi Khan, Nawab of Bengal, ceded the revenues of part of Orissa to Raghūjī Bhonsle, a Maratha leader. As a variant, in 1752 and 1760, the Nizam of Hyderabad ceded territories to the Marathas, and, in 1757, the Mughal Emperor's daughter married the son of Ahmad Shah and was granted Punjab and Sind as her dowry. As a reminder of the significance of logistics and the extent to which logistical factors posed a question-mark to conventional notions of military progress, the difficulty of financing armies ensured that much campaigning in India became an attempt to collect tribute; a problem accentuated when armies grew in size.

Cavalry did not automatically win, and Nader had supply problems, as when he operated unsuccessfully against the rebellious Lezges in Daghestan in the eastern Caucasus. Usually Nader planned logistics carefully to avoid his men running short, but on this occasion the Russians, seeing him dangerously close to their own outposts south of Astrakhan, withheld much-needed supplies of food and clothing, and prevented others from carrying them across the Caspian. Moreover, Nader's wars placed a terrible burden on his subjects and encouraged repeated opposition.[15]

Nader was assassinated by discontented officers in 1747, but there was then no resumption of Turkish expansion against Persia.[16] Baghdad was 1,334 miles from Constantinople, compared with Belgrade's 587, and there was no sea and river route to Iraq comparable to the Black Sea and the Danube and Dniester rivers to ease campaigning and logistics. That situation did not dictate an outcome, as Alexander the Great's advance indicated, but it was a factor in a situation in which the Turks faced tribal opponents in the mid-eighteenth century, but had no major conflict between the end of war with Persia in 1746 and the start of that with Russia in 1768. Moreover, military supply networks were not maintained at a level adequate to permit an easy resumption of wartime activity, which was to cause a major problem in 1768.[17]

The British had more continuous campaigning in India than the Turks did as a whole in the second half of the eighteenth century, but that did not necessarily ensure success, notably in 1779 when an expeditionary force from Bombay (Mumbai) taking its supplies with it was surrounded by Maratha cavalry and forced to surrender at Wadgaon. The crucial equations of mobility, logistics and terrain had worked against the British despite their resources. Further south in 1780, when Haidar Ali of Mysore anew invaded the Carnatic, a fertile region, his capture of magazines in the Nawab's hands, as well as the devastation spread by his cavalry, created serious supply problems for the British army, problems exacerbated by its size. As a result, the maintenance and protection of its supplies became the major operational objective of the army,[18] which helped accentuate its vulnerability. The British were able to move resources from Bengal to the troops fighting Mysore, but that drove up the burden of the struggle. Alongside the strengths of the British system, moreover, came the contingencies of particular campaigns, as well as the crucial primacy of political factors, especially the lack of unity among the Indian princes, notably conflict between Mysore and the Marathas, as in 1784–7, as well as the willingness of the Nizam of Hyderabad to ally with Britain.

The British came in India to apply power in a systematic fashion, and theirs was not an army that dispersed in order to forage or to pursue booty. Appointed Governor-General and Commander-in-Chief in India in late 1790, Charles, Viscount Cornwallis (the commander defeated by the Americans at Yorktown in 1781) swiftly wrote to General Sir William Medows, who had failed to defeat Tipu Sultan (Haidar's son) of Mysore earlier that year:

> I conceive that we can only be said to be as nearly independent of contingencies, as can be expected in war, when we are possessed of a complete battering train, and can move it with the army; and whilst we carry a large stock of provisions with us, that ample magazines shall be lodged in strong places in our rear and at no great distance from the scene of our intended operations... I hope that by a systematic activity and vigour, we shall be able to obtain decided advantage over our enemy before the commencement of the ensuing rains.[19]

Focused on the need for artillery, Cornwallis in 1791 was determined to combine the firepower that was necessary against Mysore's fortresses with a reasonable degree of mobility, and stressed the importance of both bullocks to move the artillery and cavalry. He wrote:

Large iron guns are certainly not convenient travellers, but I have always thought that unless we could carry a sufficient number of heavy guns and a good supply of money with us, we might be disappointed of gaining any material advantage from ascending the Ghats.[20]

Well-aware of logistical problems, Cornwallis seized the city of Bangalore to improve communications with the Carnatic, thus establishing a defensible forward base and creating a reliable supply system for an advance on the Mysore capital of Seringapatam. However, the latter was found to be well-fortified and defended, and Cornwallis, short of supplies, had to fall back on Bangalore, abandoning the cannon that, while necessary for sieges, slowed the rate of march.

As a reminder of the range of logistics, these issues were very different from the use of wayside grass for draught animals, a long-established usage that was an instance of an ability to use local resources that was not fully matched in the case of fleets, although they did take advantage of their facility to fish. Moreover, in India, the provision of draught animals posed problems, and, for a while, the British were unable to move their battering train and provisions efficiently because of their lack of understanding of the quality of bullocks necessary for military purposes. In addition, the advance in 1791 was affected by the epidemic that attacked the bullocks. As a result, it was only their capture of Tipu Sultan's breeding stock of bullocks in Mysore in 1799 that gave them more mobility.[21]

For several reasons, including distance from home (although not local) base and relative numbers, the problems the British encountered in India were greater than those facing the Russians in campaigning against the Turks. Nevertheless, logistics played a major role in the latter, and, as the Russians took the initiative, they had to overcome considerable distance and in a hostile terrain. When, in 1736, the Russians invaded Crimea, the Tatars avoided battle and the invaders, debilitated by disease and heat, short of food and water, and with the problems exacerbated by Tatar scorched-earth practices, retreated. Further invasions of Crimea in 1737 and 1738 were also unsuccessful, despite the improvement in logistical capability stemming from the creation of the Commissariat of War in 1711 and improvements in provisioning in 1724. Nevertheless, the

Russians captured the major Turkish fortress of Ochakov at the mouth of the River Dnieper in 1737, their advance supported by supplies brought by boat down the Dnieper and thence across the steppe by 28,000 carts. After the capture of Ochakov, the Tatars burnt the grass between the Bug and the Dniester rivers, hindering further Russian advances that year, while logistical problems prevented a crossing of the Dniester in 1738 and the Russians, more generally, suffered from a preference for heavy cavalry and heavy guns that reduced their mobility. However, in 1739, despite these drawbacks, the Russians successfully invaded Moldavia.

War resumed in 1768–74, with the weaknesses of the Turks' supply system exacerbated during the war by the Russian capture of Turkish bases and eventual advance south of the Danube. A serious shortage of food hit Turkish morale, and there were also problems with the availability and quality of gunpowder. Supply difficulties contributed to the loss of major Turkish bases, notably Chotin (Khotyn) on the Dniester in 1769, even though 50 million pounds of biscuit were requisitioned for that campaign, while, during the war, 8,500–9,000 oxen, 9,500–14,000 camels and 6,500–7,000 horses and mules were hired for logistical support by the state annually. In 1770, the Grand Vizier, Mehmed Emin Pasha, was unable to arrange adequate supplies or pay for his army.[22]

A cumulative process was at work, as Turkish operations traditionally relied heavily on supplies from the dependent territories of Moldavia and Wallachia, notably grain from the latter, now southern Romania. Their conquest by the Russians made it harder for the Turks to retain other territories, while also providing the Russians with additional supply possibilities and thus lessening the need for cumbersome supply trains.[23] The Turks were faced by a very different challenge to that posed in this region to the Romans by the Goths whom Emperor Theodosius had sought to accommodate in 382 by giving them lands in the empire under their own leaders. State development was very different in the case of Russia, and the possibility of cultural osmosis was absent.

Increasingly expert in the deployment of their forces, the Russians adopted more flexible means of supply intended to reduce the cumbersome baggage trains, although logistics remained a serious problem. The situation was eased in part due to a better ability to mobilise resources from near the zone of operations, a product of success in transforming Ukraine governmentally, politically, socially and, to a degree with Russian settlement, ethnically.[24]

The expansion of arable farming in Ukraine and Hungary served as a basis for enhancing and sustaining Russian and Austrian operations against the Turks, building capability, albeit without guaranteeing success. In addition, improvements permitted better operational planning, including greater use of river links to ease supply movements, a process helped by the flow southward of the rivers from Ukraine to the Black Sea. Nevertheless, aware of the logistical difficulties faced by the need to supply the army, Count Peter Rumiantsev in the war of 1768–74 grasped the need to take the initiative in order to produce results.[25] By the next war, that of 1787–92, the victorious Russians had largely solved the serious logistical problems of deploying and sustaining large forces across the steppe.

In these conflicts, as with Russian warfare in general, it is possible to point to serious drawbacks, but the key element was that these did not prevent tactical, operational or strategic activity and success. The adoption of more flexible means of supply helped to reduce the cumbersome baggage train of the field army, although logistics remained a major problem until the development of railways, not least because of the primitive nature of the empire's administrative system. Yet, alongside serious weaknesses in Russian logistics that owed much to 'economic underdevelopment' and much to a 'primitive' administrative structure, the Seven Years' War (1756–63) saw a large-scale and sustained projection, using 'makeshift devices', of Russian power against Prussia.[26] Furthermore, if foreign financial support was important to Russia, it also was to other states, including Austria.

France had been held up as a model, but the magazines developed to support her forces in the late seventeenth century generally only contained supplies able to support initial operations, which caused problems if the conflict was a lengthy one, and it proved impossible to transfer the cost by operating in foreign states, the situation under Louis XIV prefiguring that under Napoleon. Hit hard in the War of the Spanish Succession (1701–14 for Spain), which also became a Spanish civil war, the Spanish army and navy were then revitalised so as to be able to launch effective operations in the 1710s, 1730s and 1740s, and Spain's potential and intentions were serious issues for other states. Yet, as a result of a shortage of money, the army destined for Italy in 1741 faced multiple difficulties, including problems with the artillery.

The claim that supply problems made achieving strategic objectives only rarely possible[27] is not vindicated by the campaigning. Indeed, the

Western system of recruitment and logistics was crucial to the ability to implement strategic conceptions. Effectiveness was not simply a product of more developed systems of control and support, but these systems were important to the ability to discharge a range of functions. This system was a public-private partnership as an aspect of a customer-supplier relationship, such that Britain has moved from being presented as a fiscal-military state to that of a contractor state, the contracting out including aspects of its logistical system.[28] That partnership did not preclude improvement. Austria's military culture became characterised more by regulation than by improvisation, with a new transport corps created in 1771 and the supply system centralised. The burden of supplying the forces was certainly heavy as the military were a section of the community which governments needed and cared for. In the 1720s, every French soldier in Roussillon, a frontier province with Spain, received daily about 1 kilogram of bread, 500 grams each of vegetables and meat, 25 grams of fat, and 1 litre of wine or beer. From 1803, the Batavian (Dutch) Republic, a French client state, supplied each soldier daily with one pound of bread, the basic minimum that virtually all governments seem to have provided. Though pay was generally low and frequently delayed, troops were the largest group paid by Western governments.

Troops also had to be shod because soldiers marched on their feet, and footwear was crucial to endurance, health and morale. Invading England in 1745, Charles Edward Stuart ordered 6,000 pairs of shoes from the city authorities in Edinburgh, obtained several thousand fresh pairs in Preston, and demanded 6,000 from Glasgow on his retreat. His opponent, William, Duke of Cumberland, similarly had to obtain footwear for his men, and, between 30 May and 3 July 1746, Robert Finlay of Glasgow supplied 3,058 pairs of shoes. In such matters, the government forces benefited from the availability of ready money. On 4 December 1745, a moment of real concern in the campaign, when Cumberland was pressing his father, George II, for footwear, the general's secretary, Sir Everard Fawkener, asked the local JPs (Justices of the Peace) in the West Midlands to assemble as many horses as possible the next day, two shillings being offered for each that could carry an infantry soldier. By 9 December, Cumberland could transport 1,000 infantry that way.[29] The campaign saw both sides seeking to obtain food on the march. As the head of estates, the local élite were pushed into providing: arriving in Penrith on

18 November 1745, the Jacobite advance guard successfully demanded hay and oats from all but one of the local great houses. Correspondingly, on 3 December, Cumberland wrote to William, 3rd Duke of Devonshire, 'Should we come your way I hope that we shall find meat and bread for the soldiers'.[30]

Logistical problems remained a striking problem in Europe, as elsewhere. Troops had an equipment load that precluded them carrying their own rations, but moving supplies by cart exposed armies to one of the least effective areas of government, road maintenance, and thus to the weather. The French Moselle offensive of 1734 was delayed by poor roads and bad weather, the cannon becoming stuck in the mud. The French army is usually regarded as impressive, as indeed it was in the circumstances of the period, but, in 1734, French forces on the Rhine and Moselle suffered from a lack of provisions, and that in Lombardy delayed the start of the campaign because of a lack of forage.

Magazine systems were not only rarely adequate. In addition they could be disrupted by retreat. Thus, the French advance in the Austrian Netherlands (Belgium) in 1745–6 deprived the British of their magazines at Antwerp, Brussels and Mechelen. Moreover, magazines were of limited value if forces advanced rapidly, as the French and British did in Germany in 1741 and 1743 respectively.

Europe lacked a system of roads, let alone good roads. Instead, there were a small number of well-developed routes, such as those from Paris to Lyon or along the northern side of the Apennines from Bologna through Parma, and a mass of tracks of varying quality. Much depended on climate, soil and drainage. Impermeable soils, such as clay, quickly became quagmires after rain, but even routes on good soils could be hindered by poor drainage, heavy rains and snow melt. Following heavy rain in northern Germany in August 1758, Charles, 3rd Duke of Marlborough wrote, 'the foot [infantry] have marched the last day almost up to their middle in water the whole way'.[31] That might appear to have nothing to do with logistics, but such conditions, not least the chance of going into even deeper water, made it difficult to keep anything carried on the body dry, notably powder, and also affected baggage carts.

Bridges were infrequent, and many rivers were only crossed by ferry, a situation even more the case in North America. Deltaic regions were therefore particularly problematic. Moreover, wooden bridges and ferries were easy to destroy in wartime. Varying water courses and flooding were problems in, for example, northern Italy. In 1792, the Chinese invasion of Nepal was affected by the flooding of the Betrawati river as a result of the monsoon. Flooding was a particular problem in Europe during the spring thaw, a bad period for campaigning. The roads in Eastern Europe were especially bad, and this contributed to the marked slowness of the Russian army on the march. Whatever the quality of the route, it generally could not stand up to the pressures created by a moving army, and ruts excavated by heavy wagon wheels aggravated the situation even more, posing still greater issues for the movement of artillery.

The rivers were not much better. Few rivers had been in effect canalised, and many therefore suffered from variable water levels, weirs and tortuous meanders, while ships often had to be towed, never an easy operation in hilly country, and numerous rivers were only one-way routes. Waterlogged regions, such as the Danube valley in southern Hungary, the lower Ganges, and the lower Irrawaddy, were prone to serious diseases, and could be very dangerous for animals as well as humans. During the century and a half from 1660, canals were constructed in a number of countries, including France, Britain, Prussia and Russia, while Lombardy and the Low Countries possessed relatively good systems of water-borne transportation; but this was not a practical option in most of Europe. In addition, winter freeze, spring thaw and summer drought rendered most rivers at best undependable. In early March 1748, Cumberland, then commander of the British forces in the Low Countries, the key British army, reported: 'as yet the rivers are too full of ice to think of embarking the troops from England, as it will be utterly impossible for them to get up the Meuse'.

Thus, most supplies in Europe had to come by wagon, but the heavy wheeled transport necessary for supplies posed particular problems for the roads, while difficulties with the road surface lessened the already slow-moving progress of wheeled traffic, which acted as a major constraint on the speed of operations and, therefore on their range, not least due to the need for resupply. Long supply trains followed those armies which received their supplies in whole or part from bases or magazines. It was calculated in 1744 that the siege train required by the Allied army in the Austrian Netherlands would 'amount to 10,000 horses and 2,000 wagons', at a cost of £50,000 for six weeks.[32] Where a system of contributions had been arranged, by which intimidated areas provided supplies to prevent forcible foraging, troop dispositions had to maintain the basis of fear. Moreover, both supply trains and foraging necessitated the detaching of large numbers of troops for protection.

Aside from the military consequences, it was commonly the case that forces in the field, including in supposedly better-provided areas such as the Low Countries and Westphalia,[33] received inadequate supplies, which caused a variety of problems. Insufficient food could contribute to mutinous behaviour, as with the American Continental Army in 1781, while poor-quality food affected performance to an extent that the available records leave unclear. Limited preparation, storage and transportation techniques, including the absence of refrigeration, led to much food being spoiled. It was still eaten, and poor health was an obvious consequence, although the weakening of those, many already not especially strong, who were not listed as ill might have been more serious in terms of such activities as marching.

There were significant efforts to apply knowledge to the requirements of war,[34] notably mathematics, as with ballistics, and logistics came to the fore as a distinct term when the mathematical nature of the word was applied to operational planning. Yet, that was far easier in theory than practice, not least in light of limitations with resources.

Although the ineffectiveness of contemporary administration undoubtedly had serious military consequences, environmental issues were generally to the fore. One of the most serious was the extent to which land and sea operations generally took place in Europe for only half of every year, commonly from April till October. When the grass began growing, horses and other beasts of burden could be fed at the roadside. This was important as many armies, such as the Sardinian one operating in Lombardy in March 1734, lacked magazines of dry forage, which is a bulky product. Even when such magazines existed, they could become exhausted, as the British in the Austrian Netherlands discovered in 1744. Their operations were affected as a consequence, while the attempt to obtain contributions from occupied French territory was unsuccessful, because of the size of the demands. Forage was also to be a major problem for the French in the Seven Years' War.[35]

After the spring thaw (which proved a major issue for the Germans planning to invade the Soviet Union in 1941), roads and the land in general were usually reasonably firm until autumnal rains made routes impassable and filled siegeworks with water. Supplies were more plentiful in the late summer, when the harvest had been gathered in, one diplomat writing in May 1747 of the prospect of Frederick II 'attacking Bohemia or his other neighbours, when the forage is upon the ground, and the granaries full'. If troops slept in the field with little or no cover, winter operations could cause major losses through death, disease and desertion.

There are many instances of such operations. In 1600, Charles, Lord Mountjoy, the new English commander in Ireland, decided to campaign in the winter in order to disrupt Irish logistics, trying to immobilise the migrant herds of cattle which fed the Irish army. Although the Irish could live off the land, they were pitted against a powerful state, with a more sophisticated military and logistical organisation. In the winter of 1674–5, Austrian and other German forces, including Brandenburg-Prussian troops led by Frederick 'the Great Elector', fought a bitter campaign with the French in Alsace. On 27 December 1677, the Swedish fortress of Stettin fell to Brandenburg forces after a long siege, despite the harshness of the winter.

Nevertheless, winter operations could be costly, and those who pressed for them felt it necessary to defend their views. The Jacobite Sir Henry Goring, arguing in late 1726 for an invasion of England wrote:

> if it should be objected, that to send troops into a foreign country in the winter, they would be liable to great hardships, and that they would suffer very much by keeping the field in that season, the objection is easily answered ... they are going to their friends, who will receive them like their brothers, and make them welcome. They can suffer neither the cold nor hunger, for ... there are no garrisons in England, but a great number of fine open towns.[36]

When Bonnie Prince Charlie invaded England at the end of 1745 his hardy troops were not too badly affected by the weather, but, combined with a lack of food,[37] the weather did prevent Marshal George Wade from marching his regulars from Newcastle to Carlisle to cut him off as well as the French from mounting a supporting invasion of southern England, although the role of the British fleet was also important in the latter. Subsequently, in late December, Wade, ordered to march from Yorkshire to Newcastle 'as fast as the bad roads and this rigorous season will admit of', feared 'we shall reduce our army to nothing by long marches and encampments at this time of year'.[38] In turn, France offered the Jacobites what they lacked – siege artillery, regular infantry able to stand up to British regulars in a battle focused on an exchange of fire, and a secure supply base – but this element of logistics was dependent on being able to mount an invasion.

The unpredictability of the winter weather had also affected operations in Italy during the War of the Polish Succession (1733–5). In 1733–4, good weather allowed the French to cross the Alps and, in alliance with Charles Emmanuel III, king of Sardinia, ruler of Savoy-Piedmont, to overrun most of Lombardy, the roads being passable for their artillery. However, by February 1734, snow was hindering the French siege of Tortona and both they and their Spanish allies lost many troops due to winter sicknesses. The following winter, the armies remained in their winter quarters, affected by a shortage of money and forage, and by sickness, while spring rain delayed the beginning of the campaign. In late 1742, heavy rain prevented the British forces in the Austrian Netherlands (Belgium) from advancing into Germany to attack French forces, while a savage winter hit the French army in Bohemia hard. In the winter of 1743–4, both the Austrian and the Spanish armies in Italy suffered considerably from ill-health, desertion and shortages of wood and forage. In addition, rain turned roads into quagmires, hindering the Neapolitan invasion of the French-occupied Papal States in November 1798.

Bad weather had a more serious effect at sea, where, in a pattern that continued into later technologies, ships could lose their masts, rigging and cables in storms, and be driven aground. Supplies suffered, as did troops and horses. In December 1748, when British forces were returning from the Low Countries at the end of the War of Austrian Succession, the *Scarborough* transport lost a quarter of the horses. The difficulties of transporting horses safely by sea, combined with the logistical burden of supporting cavalry, helped ensure that the British used essentially infantry in North America, which also spared the need for fodder, harnesses, horseshoes and blacksmiths.

Britain had wealth, but that did not prevent serious supply problems. In 1729, the Regency Council had to discuss the shortage of gunpowder of the garrison on Minorca.[39] Encamped at Spire in the Rhineland in October 1743, Lieutenant-Colonel Ellison complained of a shortage of provisions and fodder, while Lieutenant-General Charles, 2nd Duke of Richmond, wrote to a leading British minister, Henry Pelham:

you ask me what we have been doing since Dettingen [British victory]. The answer is easy, nothing, then you'll say why nothing, to which I will answer…. Nothing was prepared such as bread, forage, hospitals etc to go on with.

Widespread illness and a shortage of heavy artillery were other reasons why Dettingen could not be followed by an invasion of Alsace. Logistical issues in North America were exacerbated by the issue of distance. Focusing on the need in Britain's North American colonies as war with France began in 1755, for a centralised provision of arms that could be readily moved, James De Lancey, Lieutenant Governor of New York urged:

the expediency of having, at all times, in this city, as being nearly the centre of the British colonies, a number of cannon and arms, and a large quantity of ammunition ready, on all occasion, to be disposed of for such services as the general His Majesty shall think fit to appoint for North America shall judge proper.[40]

In the event, reliance was placed on supplies moved across the Atlantic, but the logistical difficulty of repurposing for a new target was shown in 1758 after the British captured Louisbourg on Cape Breton Island, which surrendered on 26 July. There were hopes that the force could press on to attack Québec, but James Wolfe wrote on 9 August expressing army frustration:

I don't well know what we are doing here – with the harbour full of men of war, and transports – and the fine season stealing away, unenjoyed – I call it so because we should use it for the purposes of war. We have enemies close at hand, and others at a greater distance, that should in my mind be sought after … Our fleet, it seems, wants anchors, and cables, and provisions and pilots, pretty essential articles you will say … I am sure Abercromby wants assistance – we have it to spare.[41]

No expedition was to be mounted until the following year.

For all powers, there were contrasts between campaigns, contrasts that open up the question of whether they reflect more general changes across time and/or national variations, or whether they are much more episodic. The French invasions of Italy in 1635 and (more successfully) 1733 can be contrasted in terms of logistical provision. In 1741, the French advanced into Westphalia and Austria simultaneously in an impressive display of coordinated military power, but, as with the attacks in 1665 and 1672, on the Spanish Netherlands and the Dutch Republic respectively, these reflected the extent to which these were opening campaigns with readily-disposable resources, and, in contrast, by late 1742, the French army in Germany was in a much weaker state.

The greater size of eighteenth-century armies posed formidable logistical problems, especially once conflict had lasted for longer than about two years. Purchasing supplies in the field was made difficult by the limited surpluses of many areas, especially after a number of campaigns, and by the weakness of government credit. Contribution systems were of only limited use in addressing these problems, although they were employed extensively by the Austrians, for example after they conquered Bavaria in 1704 and 1742, and by the Russians in Poland in successive wars. Troops often responded to supply shortages by exactions from the civilian population, such as those of the Spaniards in the neutral Papal States in 1736, and areas where there was much campaigning suffered heavily.[42] In his *Essai Général de Tactique* (1772), a work, translated into English in 1781, that prefigured some of the practice of French Revolutionary and Napoleonic armies, Jacques, Count of Guibert, a French staff officer, not only stressed flexibility, movement, and enveloping manoeuvres, but also advocated living off the land, rather than relying on supplies being brought up, in order to increase the speed of operations.

That approach, however, was not new, for accounts of campaigns throw considerable doubt upon the customary portrayal of post-1648 European warfare as less damaging than hitherto to civilians. Indeed, there was much devastation, and notably so when regular forces fought irregulars, as in Spain during the War of the Spanish Succession, but not only them. Thus, the Austrians who invaded the French province of Dauphiné in 1692 ravaged the country and sacked and burnt the town of Gap. In December 1742, Jean-Jacques Amelot, the French Foreign Minister, attributed disorders committed by French forces in allied Bavaria to shortages, a similar excuse was made on behalf of Dutch troops in the allied Austrian Netherlands in 1744, while, in 1747, the neutral city of Liège was threatened with plunder unless it made payment to Austrian troops whose pay was in arrears. At the same time, generals responded to supply problems by altering their dispositions, Cumberland, the commander, explaining one such move in Germany in 1757 by 'The want of subsistence for the troops in the territory of Paderborn that was no longer to be procured there'.

Campaigning strain led to an increase in desertion, as with the French in Germany in 1742–3. The bleakness of the situation for wealthy Britain can be seen with the expeditionary force sent to allied Portugal in 1762. That July, Brigadier Frederick provided a depressing account of the logistical problems he faced on the march to Santarem, problems that in part arose from the poverty of the region. Arriving at Porto de Mugen, Frederick had found no beef or bread prepared for his troops and it proved impossible to obtain adequate supplies:

> all the bread that the magistrate said he could possibly get before they marched was two hundred small loaves which was so small a quantity it was impossible to divide amongst the men. I ordered the regiment to march the next morning at half an hour past three, but the carriages for the baggage not coming at the proper time it was past six before they began their march. It was late in the day before they got to Santarem.... the magistrates had provided no quarters for them neither was there beef or bread for the men, and ... they were fainting with the heat and want of food.

British generals complained about Portuguese supplies, especially of horses, mules, bread, forage and firewood.[43] Despite the difficulties, the British helped thwart a Bourbon invasion.

More generally, forces continued to operate. John, 4th Earl of Loudoun's regiment, garrisoning Fort Augustus in the Scottish Highlands, might be short of 1,057 pairs of shoes in 1747 (its complement was 1,450 men), but similar problems did not prevent other garrisons from continuing. Moreover, the comparative was crucial: British troops in the Low Countries died from fevers in August 1748 thanks to the 'wetness of this country, the bad stagnated ditch water we drink, the bad food ... we lie in barns and open cowhouses with little or no straw', but the Dutch (allies) and the French (enemies) were similarly affected.

So also with the American Patriots being able to continue operating, despite the burden of paying and supplying their forces proving a major problem as they fought the British in the War of Independence (1775–83). Despite the vitality of the agrarian economy, the demands of the war proved difficult to meet, while issues with credit and transportation further exacerbated the situation. A major attempt was made to develop industrial capability. In 1775, gunpowder mills were established in Hartford and Rhinebeck, armaments plants at Fredericksburg and Providence, and foundries constructed at Easton, East Bridgewater, Lancaster, Principio, Springfield and Trenton. Nathanael Greene, then Quartermaster-General of the Continental Army, argued in 1777 that an important reason for the defence of Philadelphia was the degree to which its industries were important for the army.[44] The rations established by Congress in 1775 for the rank and file were generous: one pound of beef, three quarters of a pound of pork or one pound of salt fish daily; one pound of bread or flour daily; one pint of milk daily; three pints of peas or beans weekly; one half-pint of rice or one pint of Indian meal weekly; and one quart of spruce beer per man or nine gallons of molasses for one hundred men weekly.

However, the gap between aspiration and reality that was such a characteristic feature of government was rarely wider than in the case of the condition of troops on active service. In light of the limited experience of Revolutionary leaders in logistical questions, the lack of an adequate central executive authority or centralised governmental machinery, the rivalries between the colonies and the active defence of local interests, logistical support for the army was, when vigorously managed, adequate; but, at other times, as a consequence of incompetence, in efficiency and selfishness, unsatisfactory. The resources that existed were mismanaged, the Patriots lacked an effective political organisation and system of taxation, and British control of the sea threw an excessive burden on the land transportation available to the Patriots. Moreover, although the militia served to suppress or inhibit Loyalist activity, local communities were not therefore disciplined to provide what was deemed necessary. The limited creditworthiness of Congress and issues with the states meant that the army had to live from hand to mouth. In 1777–8, Washington wintered at Valley Forge in part because he hoped that the rich Pennsylvania countryside would provide his men with food and forage, since what passed as Continental Army logistics were weak at best, and New Jersey was bare. In the event, the Commissariat broke down and supplies were very short that winter.

The following winter, in contrast, Washington was able to keep the largest American force-in-being of any winter of the war thanks to improvements in logistics with a newly-created system of magazines along the main lines of communications, Jeremiah Wadsworth making the Commissariat more effective and Nathanael Greene the Quartermaster's Department. Keen to keep his army together over winter, in order not to have to face the problem of re-enlistments, Washington benefited from the shelving of inland offensive operations by the British as they might have threatened the magazines. Nevertheless, in 1779, Washington was still affected by serious supply limitations: with the service departments hit hard by fiscal problems, Greene resigning in 1780 in anger with the politicians, and the attempt to rely on individual states proved seriously flawed. Robert Morris, who became Superintendent of Finance in 1781, improved the system of procurement by using private contracting rather than commissaries.

The extension of the war to the South caused new problems, not least as it was important to Washington to avoid angering the population. The army in the South in the summer of 1780 under Horatio Gates lacked organisation, and was short of supplies, leading it to collect food by threats and violence. Sufficient supplies did not arrive and the troops were forced to eat green corn and peaches. Greene, Gates' replacement, had to make operational decisions driven by supply problems.[45] Writing from Charlotte in 1780, Greene noted:

> Without tools we can do nothing and none are to be got in this country, not even a common felling axe. You will inquire of the Governor what steps have been taken by the Assembly to furnish the artificers and waggons required by me of the state, and press their immediate compliance. For without artificers we cannot aid the transportation.[46]

Given the concern at the highest level of command, it is not surprising that the frustration felt by many soldiers led to more direct action, especially a high rate of desertion, but also mutinies, notably by Pennsylvania and New Jersey regiments in January 1781, mutinies that both increased

British confidence and led the Patriots to seek money from France.[47] Contractual expectations were important to the views of Patriot soldiers and officers.[48] Moreover, there were consequences for the fighting. Thus, having driven the British back at Eutaw Springs on 8 September 1781, the American pursuit collapsed in disorder as the hungry troops looted the British camp, especially for food. This threw Greene's advance into disarray, allowing the British to rally and drive the Americans from the field; although the course of the battle has also been explained without such an emphasis on this factor.[49]

Although Patriot warships were unable to cut trans-Atlantic supply routes, the British themselves faced logistical issues, notably with the army in Canada badly affected by supply failures in 1778–9 which handicapped its ability to mount offensive operations.[50] Supplies were a problem for both sides, with the many British troops in Boston in 1775–6 short of food, while the Americans lacked powder. In 1777, Burgoyne's advance south from Canada to the Hudson Valley, a threat to the geographical cohesion, including of resources, of the Patriots, was hindered by a failure in supplies. As he subsequently told the House of Commons: 'It was soon found that in the situation of the transport system at that time, the army could barely be victualled from day to day, and that there was no prospect of establishing a magazine in due time'.[51] However, so also for his opponents in this campaign, Major-General Philip Schuyler noting at one stage 'we have no cartridge paper'.[52]

It was not logistics that brought success or failure to either side. Instead, it proved possible for both to mount operations. Nevertheless, although British control of the sea was not only crucial to their own logistics but also threw an excessive burden on the land transportation available to the Revolutionaries,[53] it did not ensure success. As for other periods, emphasis can vary, both in terms of the capability and limitations of logistics, and with reference to the role and impact of logistical factors.

In the South, the British forces suffered in 1781–2 from the American use of light forces harassing their garrisons and supply lines. More generally the use against supply lines of light forces, in part a steppe warfare technique also seen in Europe, was an element undercutting logistical effectiveness, as in 1744, both by the Austrians against the Prussians in Bohemia, and by peasants used by Charles Emanuel III of

Sardinia against a Franco-Spanish invasion force in Piedmont. In this, as in much else, logistics was an aspect of the full range of conflict.

The context for, and practice of, logistics also remained a product of the complexities of social structure and practice, and of the ambiguities of sovereignty and politics. Thus, in France, there was a tension between aristocratic assumptions about the noble nature of the officer corps and, on the other hand, the role of commercial interests. This tension reflected longstanding pressures,[54] and looked also to later differences over military appointment, promotion and politics, especially under the Third Republic (1870–1940).

Whatever its deficiencies, the French system, like its British counterpart, was able to send warships and troops on distant expeditions. These systems were a public/private arrangement, with administrators in Paris and London, in the seaports, and in the colonies, as well as merchant-financiers in each, for example Abraham Gradis in Bordeaux. Expanding the commitments of earlier systems, the networks of credit spanned government and the merchant-financiers. As such, they followed the same logistical practice seen across much of history, and notably of monetarised societies or at least those with taxation if only in goods.

This emphasis on continuity, however, does not capture the change in tasking represented by the greater scale of trans-oceanic conflict. Nor is there allowance for the engagement with the self-conscious language of improvement through method, and method through improvement. This has been termed the 'Military Enlightenment',[55] and, in part, a mathematical ethos and practice was important to that movement, although it also drew on the Scientific Revolution. Wherever located, a self-conscious intellectual military professionalism was an aspect of the situation, and an engagement with logistics as a planned process was an aspect of the professionalism that was sought. Yet, continuity is to the fore, for even the most advanced and wealthiest parts of the pre-industrial world continued to work within the kinds of logistical limits that had affected armies from Antiquity.

Chapter 5

From the Outbreak of the French Revolutionary Wars to the Age of Steam, 1792–1850

T he context for logistics was not unchanging in the century before the introduction of the large-scale application of steam-power in the shape of trains and steamships. Instead, economic growth provided more resources and opportunities, as did its demographic counterpart. This growth provided greater opportunities for existing military systems, and thus greater potential effectiveness. By 1813, Prussia had 100,000 regulars and 120,000 militia. Supplying such numbers required significant resources as well as an exploitation of the nexus between capitalism and government, seen for example in the Russian reliance on the Polish-Ukrainian economy, and notably Polish magnates, on supporting the eventually successful campaigns against the Turks in 1806–12.[1] Meanwhile, logistical pressures encouraged an emphasis on attack in order to maintain a quick decision.

The French Revolutionary forces of the 1790s dominate attention in this period, representing an apparent novelty in warfare, but the logistical challenge confronting the British army was greater than that facing any other army due to the variety and range of British commitments. Britain's ability to respond was in large part a consequence of its economic development – commercial, agricultural and industrial, the last in large part a consequence of the steam-driven machinery that was soon to be so significant for communications. This was a period of demographic expansion, with economic development of particular note in some states, most especially Britain. The logistical possibilities arising from this development and wealth were to emerge, and to differentiate Britain from France which acted in a more traditional imperial fashion of expropriation. Logistics as part of the larger economic landscape of military operations helps direct the focus in this chapter not to the continuity in logistics seen in China, Africa, Persia and the Ottoman empire, but to Britain, where industrialisation was becoming more significant, and for both navy and army.

Although eventually a global process, industrialisation proceeded at vastly contrasting rates in different places. This very disparity became a central feature of logistics, creating both resource strengths that served as a base for operations, and, also, favoured particular regions for operations. Linked to the latter came steep logistical gradients between favoured and less favoured regions, with problems for force projection into the latter. Yet, rather than focusing solely on change, the last was in part what had already occurred with operations by sedentary agrarian powers, such as Ming China in pastoral landscapes. Moreover, as the most successful of the nineteenth-century military powers, Britain is of interest.

Britain is also the focus of attention in this chapter because after the short Nepalese conflict of the early 1790s the Chinese were not involved in foreign conflict until that with Britain in the late 1830s, the First Opium War (1839–42), while the most significant conflicts in India involved Britain. This was part of a situation in which non-Western powers were increasingly looking to the West for military supplies. Sir Robert Ainslie, the experienced British Ambassador, reported from Constantinople in 1793 that the Turkish government 'has applied to me *as usual* for naval and military stores'. He itemised iron, tin bars, 100 anchors, 10,000 sheaves for pulleys, 10,000 okes (about 6,700 quarts) of paint, 22,300 shells, 45,000 round shot, 50,000 grapeshot, 20 brass mortars and their bases and carriages, 2,000 flintlocks with bayonets and 2,000 quintals of gunpowder.[2]

For Britain, alongside rivalry with expenditure on the Royal Navy, there were competing challenges for army activity, notably security in the British Isles, conflict with other powers in Europe, trans-oceanic conflict with European powers on land, and conflict with non-European powers. Between 1793 and 1815, the areas in which the army and navy operated included the British Isles, North America, the West Indies, South America, Cape Town, Egypt, Iberia, Italy, the Low Countries, Denmark, India (itself a vast area with very different environments), and the East Indies. There was no single organisation overseeing this variety, not only due to the inherited structures of British administration, but also because of the need for, and practice of, autonomy on many distant stations, which was especially the case with India where the East India Company played a major role and one that could entail tension with both army and navy. Yet, the Company also provided bases and stores, with its shipyard at Bombay very useful for naval repairs.

The extent of co-operation with the navy on trans-oceanic expeditions and amphibious operations was such that the value of differentiating army from navy should not be pushed too hard. So also with the inherited organisational structure. Thus, the Board of Ordnance provided gunpowder for both army and navy, and indeed also proved successful and pro-active in arming the navy.[3] The Navy's Victualling Board, which had responsibility for provisioning all overseas expeditions, was far more experienced and efficient, and less corrupt, than the Army Commissariat, and improvements in provisioning permitted longer-term deployment, which was key both for blockading French ports and for expeditionary warfare.[4] Logistics was one of the factors that enabled British success, and failures helped to explain mutinies.

As part of the range of bodies, departments and officials, who included the Paymaster General and the Secretaries of State, could clash, and civilian oversight and control repeatedly posed issues. In 1782, when Britain was at war, Charles, 3rd Duke of Richmond, Minister General of the Ordnance, writing about how best to repel a possible French attack on the major naval base of Plymouth, commented on:

> the many real difficulties that exist and prevent one's doing business with that dispatch that could be wished. I have many delays to surmount in my own office, but depending also upon others, upon the Commander in Chief who has his hands completely full and then upon a numerous Cabinet which is not the more expeditious for consisting of eleven persons who have each their own business to attend to.[5]

The key means of logistics, however, was not administrative structures, but money, which was crucial for funding activity, not only in Britain but also abroad, notably in the important Indian military labour market. Money was available for Britain and its global interests because of Britain's strong and developing transoceanic mercantile network. Unlike in the War of American Independence (1775–83), when, as far as hostilities in Europe were concerned, the army was restricted to defending Gibraltar against siege in 1779–83, there was large-scale conflict for Britain in Europe in 1793–1815: the British contested the French advance and presence on the European mainland, most obviously, although not only, in the Low Countries (1793–5, 1798–9, 1809, 1813–15), Southern Italy (1806), Iberia (1808–13), and France (1813–15). There was no inherent military need for such a policy, and certainly not so in terms of defending Britain from invasion. However, successive ministries felt it necessary to demonstrate to actual and potential allies that the British could challenge the French on land, which was crucial to coalition warfare, as in 1942–4 when allies frequently demanded a second front to divert German troops away from the Eastern Front. Equally, Britain needed such a statement of strength to ensure a bargaining place at the subsequent peace conference.

As a consequence, the percentage of defence spending devoted to land service rose from an average of 32 per cent in the peacetime years of 1784–92, to 51 per cent in 1793–1802, and 57 per cent in 1803–15,[6] although, in part, this rise reflected the limited possibilities for expanding expenditure on the navy, given the number of sailors that could be raised, the absence of a naval equivalent of the large forces in British pay, and the crises in French and Spanish naval power brought on by defeat. Naval manpower peaked at 147,000 in 1813, when Britain also had to support a naval and a transoceanic war with the United States, and was at about that level from 1809.[7] There was not much point in building very large warships at the end of the war, since there were no significant enemy warships at sea, although Britain was building small warships for trade protection, and thus, in part, logistical resilience, right up to the end. But the shortage of skilled seamen was the real issue, and naval commanders could have done with more men.[8]

There are problems with the analysis of available figures, as the army expenditure was always swelled by the inclusion of the subsidies transferred to Continental powers, which has led to confusion, while most of the heavy gun ordnance expenditure went towards naval guns, so that it is difficult to calculate 'land service' expenditure. The figures for 1812 presented to Parliament in 1813, were:

Army	£24,987,362 (Continental subsidy of £5,315,528 already taken out)	
Ordnance	£4,252,409	
Navy	£20,500,339	
Total	£49,740,110	
Percentages:	Army	50.2 per cent
	Ordnance	8.5 per cent
	Navy	41.2 per cent[9]

Wartime public spending was certainly unprecedented, rising from an average annual expenditure in millions of pounds of 14.8 (1756–63) and 17.4 (1777–83), to 29.2 million in 1793–1815, a figure that was higher in the later years. These figures were even more striking given the limited inflation of the period and the degree to which liquidity was far lower than in a modern economy. Moreover, on a pattern seen since the 1690s, the substantial amounts transferred as subsidies helped enable allies such as Russia to finance warfare, more especially offensive operations.[10]

The supply requirements of the forces of Britain and her allies were considerable. In late 1805, Robert, Viscount Castlereagh, the Secretary of State for War and the Colonies in 1805–6 and 1807–9, noted that the Ordnance was to provide 10,000 muskets to the Hanoverians; three years later, his correspondence covered such items as the dispatch of 300 artillery horses to the British army in Portugal and the 'half-yearly delivery of shoes to the army at home'.[11] Vast amounts of munitions were sent to allies. At the end of 1813, Castlereagh announced in Parliament that 900,000 muskets had been sent to the Continent in that year alone. The Portuguese army was well-equipped, nearly entirely by equipment from Britain.[12]

The burden of the Peninsular War with France in Portugal and Spain was particularly notable. Supplies dispatched in 1811 included 1,130 horses at the beginning of the year, clothes for 30,000 Portuguese troops, 46,756 pairs of shoes in July and August, and two portable printing presses. The costs of the Peninsular commitment mounted from £2,778,796 in 1808, to £6,061,235 in 1810, plus another £2 million in ordnance stores and in supplies in kind. Rising costs reflected increased commitments, the dispatch of more troops, and the reestablishment of the Portuguese army; this expenditure led to pressure for victory, or for the cutting or withdrawal of British forces. Britain moved from the obligation of supporting the defence of Portugal to that of seeking to overthrow the French in Spain, which was the major theme from 1812. Obliged to fight in allied countries, and, thus, unable to requisition supplies, Arthur Wellesley, later 1st Duke of Wellington, the commander from 1809, needed hard cash, but, by 1812, his shortage of money was a serious problem: the troops had not been paid for five months. When campaigning abroad, it was necessary to pay troops and foreign suppliers in British bullion, the reserves of which fell rapidly. As a result, going off the gold standard was, like the introduction of income tax, a key element in the strengthening of the logistical context and in the strategic dimension to logistics. Due to the length of the commitment, the government faced particular difficulties in meeting Wellington's demands for funds.[13]

Resources from Bengal made this issue less serious in India where, alongside logistical deficiencies, including corruption, the army administration employed both officials, notably based in forts that served as magazines, and independent entrepreneurs, notably Indian *brinjaries* (grain merchants) who wandered round with bullocks and rice looking for armies to supply; although Mysore sought to stop them supplying the British. Finance, which permitted prompt payment,[14] was a key element, and ensured that the British army was not one that dispersed in order to forage and ravage, or a force that had to be held together by booty, and that thus, at least to a degree, dedicated itself to the strategy of pillage. Logistics therefore were a factor at the tactical, operational and strategic levels of war.

The British did not always succeed. Their failure to overcome the kingdom of Kandy in the interior of Ceylon (Sri Lanka) in 1803 owed much to logistical problems, although inhospitable terrain, disease, and guerrilla attacks were also responsible. Yet, they succeeded there in 1815 and, more generally, benefiting from institutional continuity, proved able to respond to and master the range of military environments posed by India. There were far fewer British operations and far less continuity in Africa at this stage, and these operations posed greater difficulties. Having used up its ammunition, a small British force was destroyed by a much larger Asante army in West Africa in 1824 and its governor's head became a war trophy.

Like Marlborough in the War of the Spanish Succession, Wellington employed a magazine system, as opposed to the process of requisition pursued by the French. Nevertheless, Wellington's system relied on support from the host nation, whether Portugal or Spain, as well as a Commissariat that worked for the benefit of the men, and not for the system or themselves. This required Wellington being able to hold the Commissariat's feet to the fire, which was called 'Tracing the biscuit',[15] a reference to the 'hard tack' eaten by the troops.

A persistent problem, more serious than that of personalities, was provided by the convoluted command and administrative system of the army, a system that evolved in the eighteenth century as a means to prevent the army from overextending itself in politics. The Commissariat came under the Treasury and the Commissariat General, and the latter's large host of deputies and assistants, were inevitably under pressure from Whitehall. Wellington did not seek to circumvent this, but he made it clear that what he ordered was what he required. He sacked a few Commissariat generals, and other close personal staff, before getting the men he wanted.

In the face of Treasury pressures, the Secretary for War and the Colonies could provide Wellington with help, but he was extremely busy as he also had the colonies to administer from 1801. Aside from Castlereagh, senior politicians held the rank of Secretary for War and the colonies, including Henry Dundas, William Windham, and, in 1809–12, Robert, 2nd Earl of Liverpool, before he went on to become Prime Minister. Wellington complained about Liverpool, who was succeeded in 1812 by Henry, 3rd Earl Bathurst. Bathurst's tact kept the lid on Wellington, whose complaints were beyond strident, even though successive ministers did their best to keep him supplied.[16]

There were also structural problems in the role of the Transport Board, which was the key body in the planning and execution of expeditionary warfare. Economy and efficiency were in a continuous trade off, and this affected administrative structure and process. Thus, the role of the Transport Board in planning was inadequate, because Secretaries of State for War did not consult it before major Cabinet decisions were taken. Nevertheless, once preparations were in progress, there were frequent meetings, and Castlereagh clearly understood the difficulties inherent in the transport procurement process. The Transport Board did better than it had done before 1794 when William Pitt the Younger, the Prime Minister, made it effectively independent of both the army or the navy. There was no repetition of the situation at the beginning of the War of American Independence when the different departments were bidding against each other for hire of the transports.[17] In 1793–1815, the impact of the weather, and the inability of all the departments involved to perform in harmony during the preparation phase, were often underestimated. The lack of information on future requirements was also an issue, but the Board, nevertheless, skilfully used the price mechanism to attract ships, while refusing to pay an overly-high rate. The Board came to have a reputation for efficiency and were given other tasks because of this, such as the administration of the Sick and Hurt Board. The transport agents on station incurred criticism, but there was often a failure to appreciate the difficulties they faced.[18]

These difficulties were eased when distance was lessened. Thus, military success in northern Spain in 1813 enabled the British to use the harbours there, and thereby to shorten the lines of communication that had hitherto been via Lisbon; although there were still problems in developing an effective supply system. For Britain as for other powers, distance was always easier to overcome when the army was able to operate with naval support. On expeditions, troops carried their supplies with them in store ships which provided mobility, as with the supplies for 40,000 men for eight months carried by the fleet taking a large expedition to the West Indies in 1795,[19] although the ships did not carry the wagons and draught animals that helped mobility on land. When the British landed in Egypt in 1801 they 'expected no supply from the country ... we have hitherto got water – everything else is landed from the ships'.[20] Wellington repeatedly urged other commanders in Iberia that:

> I recommend to your attention my first campaign in Portugal. I kept the sea always on my flank; the transports attended the movements of the army as a magazine; and I had at all times, and every day, a short and easy communication with them. The army, therefore, could never be distressed for provisions and stores, however limited its means of land transport; and in case of necessity it might have embarked at any point of the coast.[21]

In 1813, Wellington added, 'If anyone wishes to know the history of this war, I will tell them that it is our maritime superiority gives me the power of maintaining my army while the enemy is unable to do so'.[22]

The operational side of logistics attracts most attention, but the strategic dimension was, as in war as a whole, the most significant. The British were unique both because they had cash and because their operations required naval support, these factors ensuring very different strategic parameters to those of other powers. If these parameters might seem a long way from commissariat wagoners urging unwilling oxen forward, there was in practice an important linkage.

This was important to the need to respond in theatre. The difficulties facing the Commissary General were accentuated by the lack of a collective experience. The British army had encountered major logistical difficulties in Iberia in 1703–13 and 1762, but, by 1808, when new forces were sent, there was no relevant experience. Instead, that of operating in the Low Countries in 1793–6 and 1799 was very different. In part, Iberia posed issues of limited supplies, harsh environment, and poor road system, that were very different to those in the Low Countries, as with the complaints of Lieutenant-Colonel Guard, who was in command of the important garrison at Almeida in Portugal in 1808–9. Shortages of food, clothing and shoes led British troops to pillage, and the Spanish authorities frequently did not provide the promised supplies.[23] Nevertheless, in part, whatever the area, relations with allies were a similar problem and have remained part of the politics of logistics. Issues faced were also seen in operations in British territories, notably North America in 1754–60, 1775–83, and 1812–15, and Ireland in 1798, although, in Iberia, language proved an additional burden, while the poverty of the region posed a more acute pressure on food supplies. A key aspect of poverty was the weakness of the communications network.

There was not, however, the issue of operating in hostile territory, until Wellington moved into France, advancing to Toulouse in 1814. Even then, there was concern not to offend local sensitivities, for the British were the allies of the Bourbon cause, committed to a Bourbon restoration, and reliant on local acceptance to move from military output, in the shape of victory, to political outcome in the shape of compliance. This situation was linked to the politics of logistics in the shape of not angering local opinion. In contrast, French requisitioning, which so often meant looting, compromised support for client regimes, notably that of Joseph I in Spain, and thus posed a major additional military burden in the shape of the counter-insurgency overlap of obtaining supplies; Spanish guerrilla and regular operations hit French logistics, not least by resisting French requisition parties, notably in the Montaña of Navarre. The British were harsh in their treatment of looters, with summary hanging and flogging, both carried out in front of the unit in question in order to drive home the point, and this exemplary punishment an aspect of the disciplinary system.

In part, this discipline addressed both the politics of the situation and the nature of recruitment, but it was also a response to the more particular problem posed by the juxtaposition of supply shortages at the point of operations, where troop demands were highest, with the resource-funded availability of plentiful supplies at the depots accessible to British seaborne supplies, particularly the main ports: Lisbon and Oporto. Supply shortages were a consequence of transport problems, notably the difficulties of supply columns arriving on schedule, difficulties that were accentuated when units advanced unexpectedly, whether in direction or in speed or in both, as in 1811. Conveying instructions to non-nationals in these circumstances exacerbated the difficulties, not least the strain on the commissaries. This situation was made more difficult by the extent to which the British in Iberia did not generally advance near coastlines and usually could not rely on riverine transport. Paperwork exacerbated the strain on the commissariat, though, unlike with many powers, leaving the historian with plentiful records. This paperwork was particularly apparent in the case of operations within Europe, and less so for those in India.

Logistics can be too readily separated for analytical purposes. In reality, it was, and remains, part of a supply bundle that crucially included recruitment and maintenance, the latter encompassing care for men and horses as well as equipment. In practical terms, logistics was not really separated out, and this was even more the case given the coalition dimension of British operations and its generally external location. It is easy to emphasise the disadvantages of the British army's logistical 'system' in comparison with its advantages; but, in practice, the latter were considerable, and this was even more the case with the Royal Navy.

The British found themselves against revolutionary forces in a number of conflicts, notably the French Revolutionary War, the Haitian Revolution, and the War of 1812 with the Americans (1812–15). The last revealed serious problems for both sides, that of the Americans part of a more general problem in war-fighting that in part was a consequence of the nature of the American Revolution and its military aftermath, notably the lack of structures to integrate militia and regular forces, or to deal with the supply, pay, and financing of militia forces operating outside the country. These deficiencies greatly handicapped campaigning against British-ruled Canada and were a particular problem in ensuring that the use that could have been gained from the militia was not realised.

Logistical problems in the War of 1812 interacted with those of distance and communications. Many military goods and other supplies had to be transported, by both the Americans and the British, considerable distances from the eastern seaboard. It was not possible to obtain sufficient supplies in the relatively unpopulated frontier areas, which were richer in trees than food, and it was not possible for the Americans to transfer the burden to the Canadians by invading, for the border areas of Canada had insufficient supplies. Nor was it possible for the Americans to transport the food and *matériel* they required: aside from transport and logistical problems, the necessary administrative structure was lacking. Poor weather added to the transportation problems of American forces in 1812, a Ohio volunteer noting of the march north in Ohio in June 1812: 'A continued rain for a number of days had tended to render the roads we had to travel … uncommonly bad and almost impassable for our wagons'.[24] Three months later, President Madison argued that without cannon and supplies there was no point in attacking Detroit and invading nearby Canada.[25] The following January, William Henry Harrison, an American commander and later President, wrote of 'a most unfortunate rain, which has broken up the roads, so as to render them impassable for the artillery, although it is fixed on sleds'.[26] Supply shortages hit morale and health, and encouraged desertion.

Transport difficulties in that war helped focus importance on control of the Great Lakes, notably Lake Erie in 1813, the destruction of supplies obliging the British squadron to fight the more powerful American warships, only to be defeated. Success on the lakes forced opponents to rely on the slower prospect of movement by land, and the American naval victory on Lake Champlain in 1814 led the British to abandon the effort to move south from Canada to the Hudson valley.

In comparison, the forces of Revolutionary France, whatever the theory of their logistical organisation and ethos, had relied to a considerable extent in the 1790s on seizing food in the areas into which they advanced. Although administered in a rapidly changing fashion, the new logistics brought about by the combination of need and the opportunities posed by conquest led to the partial abandonment of the magazine system and the reliance on food depots. Lazare Carnot, 'the Organiser of Victory' a pre-Revolutionary army officer with a strong interest in engineering and mathematics, who was a key figure in military administration from 1793 to 1797 and again in 1800, advocated requisitioning and thus living off the land as the means to support the large forces he helped mobilise by introducing conscription in 1793.

However, this new system faced and created problems. In particular, much of the terrain in which the French operated had only limited agrarian surpluses, and these were soon exhausted, so that it was difficult in 1799 for France's Army of Italy to find adequate food because it had already devastated much of the countryside in 1796–7, and this exacerbated desertion.[27] Moreover, transferring the burden of war to conquered territories undermined the acceptance of French control, leading to violent responses that further helped develop a vicious cycle of exaction and also undermined the universalist and liberal aspirations of France's supporters. The cause of revolution became the practice of repression.[28] The brutal exploitation of Lombardy in 1796 led to a popular uprising that was harshly repressed, and there were also serious popular uprisings in Swabia and Franconia.

Thus, the very conduct of the army, and the organisation of its supply system, helped to weaken France politically and to make it totally reliant on military success. Military convenience, lust for loot, the practice of expropriation, ideological conviction, the political advantages of successful campaigning, and strategic opportunism, all encouraged aggressive action by the French, but were also to invite revenge. In 1815, Major William Turner, part of the British army advancing after Waterloo, wrote from near Paris: 'Every town and village is completely ransacked and pillaged by the Prussians and neither wine, spirits or bread are to be found. The whole country from frontier to Paris ... laid waste'. This was linked to a desire for revenge: 'that infernal city Paris will be attacked and no doubt pillaged for it is a debt we owe to the whole of Europe'.[29]

Under Napoleon, there was more systematisation in the organisation of the French army, notably by the Chief of Staff, Louis-Alexandre Berthier, and the Minister of War, Henri Clarke, but the corps system required talented subordinates capable of independent command. The organisational structure, in the shape of independently-operating corps, provided a mobility that Napoleon used to strategic effect, with the search for a single-campaign end to the conflict. However, the pressure for a rapid close in large part as a consequence of operational surprise also reflected Napoleon's opportunism,[30] and the character of his logistical planning, in particular his emphasis on improvisation. This was seen with the successful attack on Austria in 1805, for which there were weaknesses in the logistical structure and preparations, notably insufficient forage.[31]

Improvisation meant heavy pressure on France's allies and occupied territory to produce resources, and a stress on troops living on the country. In Western Europe, where living off the land on the march was easy and could assuage the wait for supply trains to catch up with a fast-moving army, things more or less worked. In Lithuania in 1807, Spain at any point, and 1812 in Russia, the system fell apart, as roads were non-existent or poor, and the land was barren. By contrast, 1808–9 was, as part of an effective use of operational art, a logistical triumph. The Forces were moved from central Germany to Spain in 1808 almost seamlessly, in the context of the time. In 1809, when part of the army had to shift from Spain to France and then became part of a new army assembled to operate against Austria and moved to the upper Danube, this was an even greater triumph, as the numbers involved were large. Napoleon complained a lot, but did apologise to Clarke as a result of the redeployment working well, and the European powers were taken unawares by this logistical capability. Moreover, once arrived the French operated with greater resolution than the Austrians, not least with the crossing of the Danube.[32]

More generally, corruption was always endemic in the War Ministry, although Napoleon sought to curb it until he was away for too long in 1812. Other problems included muddle at the local level behind the lines, notably crossed wires and disputes with local administrators. The rhetoric of reform did not generally match the reality of a new conservatism in the shape of a highly personalised imperial monarchy bringing forward an imperial aristocracy based primarily on military service.

Napoleon's grand logistical strategy of the Continental System, a blockade of Britain by means of trade denial, was more damaging to France's allies than it was to Britain, and also made it harder both to achieve consensus in Continental Europe[33] and to negotiate any agreement with Britain. The regulations overlapped with the logistical pressure to provide for Napoleonic warfare, helping to broaden the rejection of both. Fraud was an aspect of a more general non-compliance. Even in already-conquered areas, such as Italy, the demands of military-support, whether direct, as in the requisitioning of mules, or indirect, as in higher taxation, encouraged obstruction, if not resistance. Opposition to conscription was an aspect of this wider reaction.[34]

The strategy of the Continental System helped lead to the expansion of military activity, while Napoleon's logistical method (it was scarcely a system) encouraged a movement into fresh pastures, but was no solution where supplies were limited, as in Russia and Spain, or already exhausted by frequent campaigning, and was also of limited value when French forces rested on the defensive.

In 1812, despite the scale of the logistical preparations, notably ammunition depots and supply trains with 40 days' supply for the invasion,[35] grave logistical deficiencies when the French invaded Russia, including a lack of the smaller carts necessary for the inadequate roads, combined with Russian scorched earth and guerrilla activity. The French both lost men as a consequence and, more seriously, amidst the autumnal rain, lost operational effectiveness and strategic point. Logistics became a matter of eating the horses; and retreating from Moscow the way they had advanced made the situation even more serious for the French. In his chapter on 'Maintenance and Supply' in *On War*, Clausewitz noted:

How vast a difference there is between a supply line stretching from Vilna to Moscow, where every wagon has to be procured by force, and a line from Cologne to Paris, via Liège, Louvain, Brussels, Mons, Valenciennes and Cambrai, where a commercial transaction, a bill of exchange, is enough to produce millions of rations![36]

Clausewitz thereby drew attention to the significance for the logistical landscape of pre-existing mercantile networks and agrarian wealth.

In Spain, as in Russia, resistance greatly harmed French communications and logistics, which, again contributed to the unstable character of the French hegemony. Pushed onto the defensive in Germany, Napoleon in 1813 was affected not by geography or infrastructure but by having few draught horses and oxen, and those mostly of poor quality. French supply routes were also hit by attacks from German opponents and Russian light cavalry. In 1814, fighting in France itself, Napoleon was affected by a serious shortage of arms and equipment, as well as chaos in the Ministry of War. Having sought to benefit from chance and the creation of the options of chance, Napoleon had had military and political chance increasingly imposed on him, while his options were removed.

The size of armies decreased significantly due to post-war demobilisation from 1815, although the logistical burden of the Allied occupation of much of France in 1815–18 was considerable. Smaller armies did not end logistical pressures nor the possibilities of delivering verdicts. Yet, there was no redundancy of logistical practice and impossibility of military outcome requiring a technological revolution, or, at least, change. Instead, alongside, and as part of, existing systems, the pressure of campaign produced expedients in the usual manner of fitness-for-purpose.

This was abundantly clear in the Latin American Wars of Independence, in which both sides faced significant problems. Royalist forces were obliged to rely on forced loans and seizing local supplies, which eroded support, but revolutionary forces, notably in Venezuela, were short of pay and supplies. Thanks to fewer resources, worse communications, and a larger area of operations, logistics were a greater problem than in Europe, the raising of supplies through force was normal, and there was much burning and destruction of crop, *haciendas* and towns, in order both to deny resources and to punish.

This situation remained the case in subsequent conflict in Latin America which was one of the most violent parts of the world for the remainder of the century and where warmaking states could well get weaker. The premium was on small forces that could take their supplies with them and/or seize them en route. Logistical problems were particularly the case in civil wars. Thus, in Brazil, in facing the *Cabanos* revolt in the Pernambuco region in 1832–5, government forces were hindered by the size of the areas of operation, poor communications, and a lack of adequate pay and food which led to desertion. The situation was exacerbated by the disruption caused by the conflict and by *Cabano* guerrillas. However, in 1835, the government forces became more active, destroying *Cabano* crops in forest regions, isolating the *Cabanos*, and leaving them increasingly short of food.[37] Similarly, in Yucatán in 1847–8, Mexican troops in the so-called Caste War seized goods crucial to rebel logistics and destroyed their settlements. In operations against Native Americans, there was the destruction of crops and food stores, the seizure of horses, and the killing of bison, a practice also used by Spain against the Apache in the 1790s. The significance of control over food ensured that, in the face of the Taiping Rebellion of 1850–64 in China, the largest civil war in the nineteenth-century world, the Manchu government was particularly keen to maintain control of areas producing grain surpluses, which forced the Taiping to go on, in effect, raiding expeditions.[38]

In contrast, Winfield Scott, the brilliant American commander in Mexico in 1847–8, followed a strategy of conciliation, at least to the élites. Aware of the danger of a guerrilla rising, he avoided relying on the logistics of depredation.[39] In his successful advance on Mexico City in 1847, Scott followed a pattern that was to continue to be seen in the second half of the century, cutting away from his lines of communication and pressing on despite being in hostile country. In part this was due to the government's decision that the initial invasion, from the north by Zachary Taylor, faced too long a route to Mexico City, with consequences in terms of opposition and logistics. At the same time, Scott's success was not simply because of his skill combined with a shorter route, but also owed much to the problems facing the Mexicans which included political division, and a shortage of funds and supplies.[40] In contrast, American logistics ultimately rested on fiscal strength in the shape of customs revenue and borrowing capacity. Lacking this political cohesion and public credit, Mexico relied on expropriation. Fiscal-military capacity ultimately depended on political culture.

Conflict in Iberia after 1815 had many similarities with civil wars in Latin America, with forces frequently affected by a shortage of pay and supplies, as in the First Carlist War in Spain in 1832–40. The extensive role of adventurers, not least American filibusterers in Central America, also ensured that much logistics was *ad hoc* and for specific tasks.[41]

These points deserve as much attention as the increase in capability that tends to be cited, for example the change in Britain from the naval bakery built in Portsea in 1724 to the new, larger factory opened in 1828 in Gosport, again near the naval base of Portsmouth, with mass production equipment that was capable of producing 10,000 naval biscuits daily.[42]

Differing outcomes were in part a matter of resources, but tasking was more significant, which underlined the role of strategic culture. In the Opium Wars with China (1839–42, 1856–60), the British (and the French also in the Second) limited their military activities and therefore kept themselves from straining their logistical system: there was no deep advance into China comparable to that which proved Japan's undoing in 1937–45. So also with the impact of tasking on weaponry, which, in turn, had specific logistical requirements. Thus, the mobility required of artillery could vary depending on how far a state was interested in expansion.

There was also the increasingly systematised process of military education for officers. This entailed consideration of recent conflicts, and of the texts accordingly written. The engagement with Napoleon's use of envelopment as part of a war of manoeuvre encouraged an interest in lines of communication, and supply routes were an integral part of the latter.[43]

In the conceiving of 'lessons', a key need was not to treat logistics (or any other aspects of warfare) as a uniform factor, but, rather, to accept that a prime need was to adapt structures to very different environments. Thus, Napoleon failed, both in Poland in 1806–7 and in Russia in 1812, to respond to the particular problems of the campaigning zone, notably a lack of locally-available food and poor roads.[44] The ability to understand circumstances and to be flexible were the key logistical requirements. Systems themselves did not provide this outcome. It is striking that Napoleon's system was praised so strongly by van Creveld, but that detailed studies have been more critical. The French policy of obtaining food locally, of war feeding war, repeatedly proved inadequate and a cause of opposition, while armies found their logistical support inadequate.[45] These problems underlined the systemic crisis of Napoleonic warfare, one that is underplayed due to the focus on his successful campaigns. A consideration of Napoleon, indeed, poses wider questions about the nature of writing on logistics, strategy and warfare as a whole.

More specifically, the period covered by this chapter acts as a hinge between the earlier chapters, dealing with logistical possibilities in the context of long-term agrarian continuity and therefore agrarian-based logistics, and the following ones addressing rapid, subsequent change.

Chapter 6

From the Age of Steam to the First World War, 1850–1914

'thanks to coal, and thanks to coke
We never run a ship ashore!'
Captain Sir Edward Corcoran R.N.
from Gilbert and Sullivan's *Utopia Limited* (1893), Act 1.

Military activity in the early nineteenth century showed much of centuries-old characteristics, and logistics was part of this, whether in terms of the nature and movement of supplies or of the assumptions and consequences linked to them. A century later, the situation was very different, and therefore for this chapter the emphasis is very much on change, with technological innovation and, crucially, diffusion and implementation linked to the opportunities and needs created by rapidly growing populations and economies. The focus of particular attention is Britain, as the global power that drew heavily on steam power, but also the United States, because of the role of rail in the logistics of the Civil War and due to its future significance.

Although they had begun earlier in the century, the new technologies of transport by land and sea, railways and steamships, became in the mid-nineteenth century more established and more integrated into the operations of armies and navies; although, at the same time, these technologies both had vulnerabilities and posed their own logistical requirements. Steam became the key motive power, one that was even more important at sea than on land, and logistics was transformed as part of an organisational-industrial-technological nexus that was inherent to the process of change in this period. The organisational dimension itself had many aspects, so that, for example, exports for armaments were important to the ability to build up state production.

For the military, the transformation of challenges and possibilities was compounded by the sustained demographic expansion that began in the mid-eighteenth century, ensuring that there were more young men, a key feature of the French Revolutionary and Napoleonic Wars (1792–1815), and that the military could raise greater numbers without restricting the labour force at a time when economic demands for labour were rising significantly. The world population rose from about one billion in 1800 to about 1.6 in 1900. Alongside the adoption of new weaponry, there were important changes in organisation that had consequences for logistics, notably top-down control of the military-industrial establishment, improved officer education, and a determination to improve the handling of logistics and communications. Napoleon's triumphs and failures helped clarify what was operationally necessary, and were seen in that light. The development of the operational level of war in late nineteenth-century Europe, was both cause and effect of the development of what the British called the administrative staff, intertwined with the army at most levels of organisations, and, separately, of the General Staff which was based on a Prussian/German model.

Despite important improvements in capacity and running, railways, once introduced, lacked a comparable process of enhancement to that of steamships, but they proved highly significant in war. By the Crimean War (1854–6), they were playing a role in warfare, in this case largely because the lack of Russian rail links to Crimea hit the deployment and logistics of Russian forces there. Thus, due to the weakness of its rail system, Russia was unable to realise much of the potential of its central position when attacked around its perimeter by seaborne Anglo-French forces able to take advantage of the capability offered by maritime steam power. More generally, the Allies' lead in weaponry rested on a superior technical and manufacturing base.[1] The war can be seen for Britain in terms of failure and success. The troops sent to Crimea in 1854 had insufficient tents, wagons and medical support. However, the consolidation of the War Office, Horse Guards, and Board of Ordnance, a key administrative development, began in 1854–5, and there were improvements in the handling of logistics. The Royal Wagon Corps established in 1799 but disbanded in 1833 was refounded as the Land Transport Corps in 1855 and in 1856 this became the Military Train which became a permanent, rather than just wartime, body.[2] Moreover, a formidable quantity of supplies was delivered: in the siege of Sevastopol in 1854–5, the Allies fired 1,350,000 rounds of artillery ammunition.

In contrast to the Crimean War, in the Austro-French war in far closer Italy in 1859, both sides employed railways in the mobilisation and deployment of their forces. In the opening stage of the conflict, despite problems with only single-track lines and insufficient locomotives on the part of their ally Piedmont, the French moved 130,000 troops to Italy by rail in a matter of weeks, thereby helping to gain the initiative. This number includes those transported within France to Marseille and Toulon for subsequent movement by sea to Genoa (then part of Piedmont),[3] a process that was faster and more reliable than the movement of Charles VIII's artillery in 1494.

The extent to which the railway had fundamentally altered the relationship between amphibious attack and land defence was to be discussed by geopoliticians in the early twentieth century, notably Halford Mackinder in 1904,[4] but was already playing a role in military planning by the mid-nineteenth. In 1854, Sir James Graham, the First Lord of the British Admiralty, wrote to Fitzroy, Lord Raglan, the Master-General of the Ordnance, about the need to defend the estuary of the Humber on Britain's east coast: 'I quite concur in the opinion that the permanent presence of a large military force at Hull [the major port] is not requisite: that inland concentration, with rapid means of distribution by railroad is the right system'.[5]

Military plans increasingly depended on using or threatening rail links. In 1861, Captain William Noble of the British Royal Engineers produced a report on the defence of Canada (part of the British empire) arguing that the American presence on the St Lawrence waterway was a threat to British communications, as the only railway from Montréal to Kingston, Ontario and points further west went along the north bank of the river. As a result, Noble pressed for Britain taking the initiative if war broke out with America.[6]

In North America and Europe, the 1860s was to be a key decade in the military use of railways. In the American Civil War (1861–5), the railway, which had expanded greatly in America in the 1840s and 1850s, especially in the North, made a huge difference, tactically, operationally, strategically, and economically. At the tactical level, man-made landscape features created for railways, such as embankments, played a part in battles.

Operationally, the railway created links along which troops could move. In 1862, the Confederate (South's) commander Braxton Bragg was able to move his troops 776 miles by rail from Mississippi to Chattanooga, and thus create an opportunity for an invasion of Kentucky. Such a potential was totally different to the situation during the previous wars in North America, and, if the term revolutionary is helpful, then the capacity to plan for rapid movement was a significant development.[7]

As a result, rail junctions or river ports where steamship services and railways were linked, such as Atlanta, Chattanooga, Corinth, Manassas and Nashville, became operationally highly significant and the object, or at least focus, of campaigning. From the outset, Union (North) advances aimed at such junctions, which the Confederates struggled to protect or regain. Moreover, plans were discussed in these terms. Richard, Lord Lyons, the British envoy, reported in February 1862 on Union moves and options west of the Appalachians:

> The fall of Fort Donelson and Fort Henry has given the Federals the command of the Cumberland and Tennessee Rivers – They to be thus enabled to occupy the Western part of Tennessee, to obtain possession of Nashville and the railroads which united at that point, and in this way to interrupt the communication between Virginia and the South through Tennessee.[8]

Indeed, west of the Appalachians, the Union forces planned both to gain control of the Mississippi River and to advance into the Confederacy along the railways running southeast through Tennessee and Georgia.

In 1862, in the campaign east of the Appalachians that led to the Battle of Second Manassas or Bull Run, the Confederate general 'Stonewall' Jackson hit the Union supply route along the Orange and Alexandria Railroad, destroying the supply depot at Manassas Junction. Later that year, the Union commander Ambrose Burnside planned to move south toward the Confederate capital, Richmond, along the Richmond, Fredericksburg and Potomac Railway after he had captured the river crossing point of Fredericksburg, which, in the event, he failed to do. In 1864, the Union failure to cut the Richmond and Petersburg Railroad meant that the Confederate Army of Northern Virginia retained its major supply line, but the success in cutting rail links led the Confederates to abandon Atlanta. This victory helped ensure Abraham

Lincoln's re-election which had looked less probable earlier in the year. The comparable campaign in the War of American Independence (itself a civil war) had been the British surrounding of Charleston in 1780, but that had been obtained, thanks to amphibious capability, by moving units across rivers by boat, notably the Ashley and the Cooper. The Patriots had then held on in Charleston, only to surrender rapidly once it was bombarded.

The use and potential of technology combined in the Civil War to ensure organisational attention. Building on the already strong relationship between the army and civil engineering, notably the railways,[9] the Union created US Military Railroads as a branch of the War Department. This was part of the process of wartime mobilisation, the scope of which expanded as a result of the organisational demands stemming from new capacity and needs. The function of US Military Railroads included the building and repair of track and bridges. This capability provided a quick-response system to such requirements for transport links as the exigencies of the war required. As with the purchasing networks and naval dockyards created in the early-modern period to sustain warships, this system was necessary in order to give effect to the new technology, as well as to earlier systems operating at the same time and in concert, notably wagon-based logistics.

Alongside the weaknesses of the Confederacy's railway infrastructure,[10] it is important to put the use of the railway in the Civil War in context. Railheads were significant to campaigning, but the rail network did not necessarily influence where the campaigns were fought. Rivers, not least because of the valleys, and thus corridor, through mountain ranges, they provided,[11] probably had a greater effect on where operations were mounted, although William Tecumseh Sherman and Ulysses S. Grant's remorseless offensives in 1864–5 moved away from the shackles of a confined geography which rivers and, indeed, railways imposed. This was a manifestation of the change in the nature of conflict during the war. Indeed, there were no railways in the Wilderness in Virginia, and it was here in 1864 perhaps more than anywhere that the novelty of trench warfare made an impact.

There was still a heavy dependence for logistics in this war on such traditional means as wagons, soldiers carrying their own supplies, and foraging. These were important, not least because much of the terrain was wooded and with a low density of roads and tracks. This situation helped ensure that incremental, rather than revolutionary, improvement, for example in field transport, was a key element,[12] and notably to Sherman's successful advance through Georgia and the Carolinas in 1864–5, an advance that destroyed the Confederacy's strategic depth. Sherman was supported by five wagon trains with a total of 2,500 wagons, and nearly a third of his force was used to protect them. The advance thus entailed the commonplace problem of trying to keep troops and supplies in harness, a problem exacerbated by the state of the roads.[13] They were not designed for such a capacity, which, in turn, degraded the roads. The contrast between railways and wagon-dependent moves indicates the degree to which the strategic level of war was transformed more at this stage by industrialisation than operations, and the latter more than its tactical counterpart. This introduced a tension between new strategic possibilities and still limited operational and, even more, tactical means to implement them. Like much else, this tension was presaged in the Civil War but came to a full flower in the First World War.

From another perspective, and as with communications and logistics in other conflicts, the problem with drawing an overall conclusion for the Civil War is that, on the one hand, railways seemed to influence where campaigns and battles were fought, while others were dictated by the presence of navigable rivers; but some campaigns paid little or no direct heed to the presence of either. Union forces found difficulties in advancing into the Confederacy, not least due to the deficiencies of the roads and the weaknesses of the rail system, but their absence or presence were not the sole factors.[14]

Moreover, the technological aspects of railway warfare were limited. There were no major developments in the specifications of locomotives. Rail design and fabrication methods, as well as railway construction systems, were crucial to the speed of laying of new track in addition to the safety of trains on it and the speed at which they could travel and thus move supplies. Changes in these factors were not driven forward by the Civil War, and economic developments, rather than the war, were responsible for the growth of a large-volume rail system. So also with steamboat technology. Most wars of the period were too short to produce at the time any major changes in technology.

The concept of fitness for purpose helps explain the degree to which there was improvement in specific conflicts, whatever their character. As with other struggles in which the public as a whole were mobilised, the Civil War was a prime example, and repays examination not, as earlier in the chapter, of the transport methods available for logistics, but, rather, of the resource and organisational means. There was a major contrast in this respect because of the greater resources of the Union. It had a edge in manpower (one increased by the racist policy of the Confederacy which precluded slave soldiers) as well as formidable advantage in manufacturing plant, railway track and bullion resources, the gold and silver of the West being especially important. The purchase of government bonds helped provide finance, as did an extension of taxation.[15] The disparity was accentuated by the economic and financial dislocation stemming from the Union blockade, and by the systemic weakness of a Confederate central government saddled with the incubus of 'states' rights'.

Union resources made it easier to equip the large numbers of troops raised, as well as a significant navy.[16] Thus, nearly 1.5 million Springfield rifles were manufactured, a total which reflected the capability of contemporary industry, and one which could not be matched in the Confederacy. The Union's agricultural strength in part rested on an ability to respond to new possibilities, especially agricultural machinery. As a consequence, wheat output rose in 1861 and 1862, even though part of the agricultural workforce served in the army. Supply problems in the Confederacy were far more serious.[17] They encouraged Lee to march into the North in 1862 and 1863: he hoped to gain food, boots and other supplies, shortages of which were affecting morale and effectiveness. Moreover, clearing the Shenandoah Valley offered the possibility of securing Virginia's grain for his army, while operating in Pennsylvania would lessen the burden on Virginia.[18] In addition, the capture of the Pennsylvania coalfield in 1863 would have hit the Union's industry and logistics, with rail totally dependent on coal. In turn, the impact of logistical problems, notably the supply of fodder, played a role in the Union's failure to prevent Lee's successful retreat after Gettysburg.[19] Lee's invasion of the North had major political implications: it helped galvanise support for the war in the Union as the Confederacy now looked like an aggressor, which also had implications on the global stage.

Aside from resource advantages, the Union increasingly organised and systematised their logistics, with organisation a key factor, and the Quartermaster Department, established in 1818, being greatly expanded to include over 100,000 civilian employees. It manufactured or obtained formidable amounts, including six million woollen blankets and one million horses and mules. The relationships between the federal level, and both state quartermasters and private businesses were often antagonistic, but not, as in the South, crippling. There was no doubt about the scale of Union effort. This was indicated by the sale in Fiscal Year 1866 by the Quartermaster Department of 12,534 regular wagons, 3,357 other wagons, and 820 movable forges. 83,887 civilian wage employees were discharged by the close of September 1865.[20] More generally, the demand for *materiél* increased considerably. In 1864, Prussia deployed 122 rifled cannon at the siege of Dybbøl in Denmark and by the close of the siege they were firing at least 4,000 shells and maybe twice as much daily, shells brought by train and wagon.

Attempts at logistical improvement in the Civil War faced the degree to which the new potential for transport and control created the problem of a new standard of achievement. This problem was exacerbated by the speed with which the new systems had to be brought into play and over a good area. Neither side can be described as war machines if that is intended to suggest predictable and regular operating systems which could be readily controlled and adapted. There was a shortage of effective staff structures and command systems. Organisation at the centre was not matched, and certainly not in the early years of the war, by counterparts in individual armies on campaign.[21]

At the same time, logistical issues did not pertain in a values-free environment. Thus, in 1862, and more particularly in 1863, there was the development of 'hard war,' with Union forces responding harshly to opposition by becoming more destructive and especially by living off the land and controlling local food supplies.[22] This was an aspect of a more general abandonment of conciliation toward the Confederacy,[23] an abandonment that reflected the failure to bring the war to a rapid close. Thus, logistics was part of the psychological mastery of Southern society that was designed to lead to the collapse of the Confederacy.[24]

On a different scale, in space and time terms, to the American Civil War, an effective quick-response transport and logistics system was also created by Prussia in the 1860s. In its wars with Austria in 1866 and France in 1870–1, Prussia won due to a combination of different factors: the mistakes of its opponents, the combat quality of Prussian officers and troops, and thanks to an effective exploitation of the railway network to achieve rapidly the desired initial deployment, thereby gaining the operational and strategic initiative. The Crimean and Franco-Austrian wars had led to an understanding in Europe of the need for good rail links for military purposes, and, as this was something that could be done, it was. Through use of planned rail movements, the Prussians made mobilisation a rapid and predictable sequence and greatly eased the concentration of forces. Uniform speed by all trains was employed by the Prussians in order to maximise their use, and military trains were also of standard length. Rail use was planned and controlled by a railroad commission and by line commands, creating an integrated system that was linked and made responsive by the telegraph, an instance of the multiplication of capability by means of the integration of technologies, and also of the reliance of logistics on information and its rapid dissemination. The number of locomotives was kept high in order to cope with a wartime rise in demand. The key point, which Moltke seems to have understood and carried out intentionally, was to make mobilisation followed by concentration and then invasion part of the same seamless movement, ideally all the way through to the final battle (*Schlacht ohn Morgen*). This capability took on greater value given the response. Although the French had a better railway system, they made worse use of it. The French tried to combine mobilisation and concentration, and created chaos. From the early frontier battles through to that of Sedan, French units were crippled or paralysed by logistical problems.[25] Yet, in 1914, once the fighting began, it was to be the French use of rail in order to move troops that proved more effective than that of Germany. Being on the defensive, the French could use their own rail network, as the Soviets were to do in 1941.

The availability and use of railways and the associated infrastructure helped both to create capability gaps or gradients within the West, and also, more obviously, between the West and non-Western powers with the railways very important in imperial expansion. In the American Civil War, regiments were deployed from the train onto the battlefield, as at Chickamauga, but because railways were usually built behind the front of imperial advance, they were not generally involved in conflict in a tactical sense, where the limited mobility of troops and supplies was still very striking, or, usually, operationally. This remained the case into the twentieth century, for example with the Italian conquest of Ethiopia in 1935–6. Nevertheless, railways could be of operational significance for advancing forces if the pace of advance was reasonably slow, and they played a key role in logistics, a factor that was of greater importance as Western forces operated inland, for, in the absence of river routes, such operations compromised the power-projection advantage of ocean steamers.

Moreover, the need to use human porters for many inland operations created problems in the shape of obtaining sufficient porters and then providing them with enough food. Native bearers were important in sub-Saharan conflict, for example to the well-planned British advance on Kumasi, the Asante capital, in 1874.[26] The Asante attitude before the war owed much to their belief that the British would not be able to overcome logistical problems to reach Kumasi. These bearers were not slave soldiers,[27] but were a variant on that long-standing practice of the African military labour market. The high death-rate of porters was a significant operational element, notably in the African campaigns of the First World War in which the German colonies were conquered. Where possible, there was across the world also a continued role for horses, mules and oxen, which, together, provided essential mobility for long-range movements away from rail links.

Nevertheless, in 1896, the Anglo-Egyptian army invading Sudan built a railway straight across the desert from Wadi Halfa to Abu Hamed, the 383-mile-long Sudan Military Railway constructed across the bend of the Nile, and pushed onto Atbara in 1898, playing a major role in the supply of the British forces, and thus in operational effectiveness. The British crushed the Sudanese forces at Omdurman in 1898. Steamers on the Nile were also part of their logistical system, and, moreover, were easier to protect than rail track. In the Second Boer War (1899–1902), the railways that ran inland from the South African ports such as Durban and Cape Town facilitated the deployment of British military resources, first against Boer advances and, subsequently, in the conquest of the Orange Free State and the Transvaal; although wagon trains also proved very necessary.

The British also needed horses for mobility to win, and bought and transported about half a million, mostly by ship, from all over the world. However, they could not acclimatise them or indeed feed them properly for most of the war, and about two-thirds of them died of starvation and misuse, being treated as disposable.[28]

Railways were regarded as of strategic importance, in particular anchoring territorial interests, and their extension was a matter of governmental concern and press comment. By 1906, a rail system in Russian Central Asia had been created, serving strategic and economic interests and troubling British observers worried about a threat to Afghanistan and Persia (Iran), and hence to the British position in southern Asia. Moreover, to the east, in 1896 Russia obliged China to grant a concession for a railway across Manchuria to Vladivostok, and this Chinese Eastern Railway (CER) was constructed in 1897–1904. When the Russians advanced against the Chinese in Manchuria in 1900, the railway served as an axis of movement. The CER proved a key issue of Russian power until purchased by Japan in 1934.[29] The scope of the Russian rail system again became an issue during war with Japan in 1904–5, in this case with the need to reinforce and sustain, across the vast distance of Siberia, the peripheral but strategically significant position in Manchuria against Japanese attack. Although, in 1904, the Trans-Siberian Railway was incomplete at Lake Baikal, the Russians transported 370,000 troops along it to the east. Their ability to use the railways as an integrated element of strategy making, however, has been queried.[30]

Communication links were sought by imperial powers and were seen as a way to strengthen their empires, politically, economically, militarily and symbolically.[31] The role of the railway therefore was strategic and logistical in the broadest sense, in that, like Roman and Inca roads and Chinese canals, it helped in the development of the economic links that sustained and strengthened the major powers. For example, the railway was crucial in creating and improving economic links between coastal and hinterland America, as well as in integrating the frontiers of settlement with the world economy, an integration that was important in the spread of cultivation and ranching, with the cattle being driven to railheads, and in the exploitation of mining opportunities. The combination of railways, steamships, credit and trade helped ensure that the 'logistical space' of most states was transformed. The need to retain control of farming regions in order to mount operations became less significant. In contrast, in China, in the face of the Taiping Rebellion of 1850–64, the Manchu government was particularly keen to maintain control of areas producing grain surpluses. This forced the Taiping to go on what in effect were raiding expeditions.[32]

Railways were important to the protection of empire. Faced by the rebellion of the *Métis* (mixed-blood population) in Manitoba and Saskatchewan in 1885, the Canadian government sent over 4,000 militia and their supplies west over the Canadian Pacific Railway, achieving an overwhelming superiority that helped bring rapid victory. This display of capability was followed by a new increase in government subsidy for the railway that enabled its completion that year.[33] Similarly, the railways the British built in India helped ensure that troops and their supplies could be moved to areas of difficulty, as in 1897 when they were sent to the North-West Frontier (of modern Pakistan) to assist in overcoming resistance among the Waziris.[34] Railheads, notably Peshawar and Quetta, played a key part in British planning both on that frontier and with regard to power-projection into Afghanistan. In turn, these railheads and the routes from them had to be protected; while an anxious eye was kept on the development of the Russian rail network. This was a major aspect of the Intelligence side of logistics, an aspect that should always be borne in mind as capabilities exist not in the abstract but with reference to particular tasks and opponents.

In Mexico, the building of a railway across the rebel area in 1900 helped to end long-standing Mayan resistance to the Mexican government, although the effects of cholera, smallpox and whooping cough were also significant. In 1926, the French responded to a rebellion in their new colony of Syria by building a railway into the Jebel Druze region, the centre of the opposition, in order to facilitate the movement of their forces.

A simpler technology than that of railways, namely roads, remained highly significant. Usage in the nineteenth century did not yet focus on vehicles equipped with internal combustion engines, but the construction and maintenance of roads required certain techniques, such as cambering, to increase their durability and thus resistance to weather and wear-and-tear, while the technological element rested with the equipment used for construction and maintenance. The high explosives that helped disrupt and move large quantities of rock, as in tunnel construction, were important. Furthermore, the relevance of roads in the late nineteenth century was not a case of a residual significance in the gaps between new railways, for many new roads were built. For example, road construction was crucial in New Zealand in the 1860s, when British troops were used to extend the Great South Road from Auckland over the hills south of Drury in the face of Maori opposition. By 1874, a road had been finished between Tauranga and Napier, separating two areas of Maori dissidence.

Steamships changed logistics. Early versions suffered from slow speed, a high rate of coal consumption, and the space taken up by side and paddle wheels, and coal bunkers. But steam power replaced dependence on the wind, making journey times more predictable and quicker, and increased the manoeuvrability of ships, making it easier to sound and sail inshore and hazardous waters, and to operate during bad weather. Moreover, maritime steam technology developed rapidly, with the screw propeller (placed at the stern) increasing mobility from the 1840s. The global range of British commitments fed into the development of strategic communications for political, military and economic purposes, with the resulting development of technology.[35] This was an important aspect of logistical capability.

The necessary extension of the network of British coaling stations made possible by the global nature of Britain's empire, ensured that its steam-powered armoured warships could be used in deep waters across the world. In late 1861, as war with the United States apparently neared, Britain, the leading producer in the world, sent coal to New World bases, notably Bermuda, to support a larger naval presence there. In 1898, even an incomplete list of British naval bases included Wellington, Fiji, Sydney, Melbourne, Adelaide, Albany, Cape York (Australia), Labuan (North Borneo), Singapore, Hong Kong, Weihaiwei (China), Calcutta, Bombay, Trincomalee, Colombo, the Seychelles, Mauritius, Zanzibar, Mombasa, Aden, Cape Town, St Helena, Ascension, Lagos, Malta, Gibraltar, Alexandria, Halifax, Bermuda, Jamaica, Antigua, St Lucia, Trinidad, Port Stanley in the Falkland Islands, and Esquimalt (British Columbia). An unwelcome reliance on British colliers encouraged the Americans to move to oil-firing steam engines in 1913.[36] France had the second largest empire in the world, and its overseas naval bases included Martinique, Guadeloupe, Oran, Bizerta, Dakar, Libreville, Diego Suarez, Obok, Saigon, Kwangchowan, New Caledonia, and Tahiti. Russia found access to coal a significant problem in moving its Baltic fleet to Japanese waters in 1904–5.

Despite interest in foreign coaling stations, the United States for long did not follow this trend, in part due to a wariness about acquiring colonies, but also because coal could be purchased.[37] In the shorter term, this lack of bases encouraged interest in methods to reduce need, notably fuel efficiency and more direct navigation; in turn, the situation encouraged a turn to oil. Britain could produce its own coal, and its change to oil locked Britain into needing control of territories for a resource that it, unlike the United States, did not produce at home.[38]

Naval bases served as nodal points for the movement of military resources – men, munitions, ships – which was such an important aspect of imperial success, permitting the concentration of resources when required and, more generally, the transfer of such resources in a planned fashion as part of a system of power. Thus, in 1837–40, the Dutch moved recruits raised on the Gold Coast in West Africa to the East Indies. The forces that could be supported increased. Whereas the Portuguese had sent 400 musketeers to the aid of Ethiopia in 1541, a long distance round Africa, the British invaded with 13,000 troops from relatively nearby India in 1867.

Alongside the standard account of logistics – all those railways, steamships and supplies – there was also, however, a continuity in terms of the problems posed in many areas for all combatants by poor logistical systems and shortages of supplies. In part, this was a matter of the incomplete penetration of various areas by networks of transport and exchange at various levels of techno-economic sophistication and was just a continuation of what had always been true.

Thus, logistical problems, notably the subsistence agriculture of the drier savannas and the problems of operating in rain forest, helped ensure that much campaigning by African forces was a matter of essentially huge raids.[39] The provision of firearms and, in particular, ammunition was important to African capability, as in Ethiopia in the 1870s. At the same time, provision of the necessary gunpowder as well as ammunition resupply were major issues.[40]

Across much of the world, size-supply ratios were, as ever, a crucial element logistically. Small forces continued to be able to take their supplies with them and/or seize them *en route*. Winfield Scott had only 10,700 effectives for his final advance on Mexico City. The Americans also used independently-operating columns with cavalry and wagon support in their operations against Native Americans, as in the Red River War of 1874–5 with the Southern Cheyenne. Effective logistics enabled the staging of winter campaigns, as in north Texas in late 1868, and this put particular pressure on the Native Americans. At the same time, the logistics offered targets to opponents, notably with attacks on the supporting supply train. Thus, operational capability had a tactical cost.

This situation was similar to many of the forces deployed by Western states during the 'Age of Imperialism', with Winston Churchill observing in 1899 of the recent British campaign in Sudan, 'Victory is the beautiful, bright coloured flower. Transport is the stem without which it could never have blossomed';[41] but such logistics was very different in scale and operational complexity from the main-battle armies of the leading powers involved in symmetrical warfare during the century. Thus, the possibilities offered by technological change depended on the tasking and the geographical context. With rail, although new track could be rapidly laid, there was the need to use existing facilities, as with the French forces brought to suppress revolution in Paris in 1848 and Rome in 1849, and into Italy to fight the Austrians in 1859. Such use suggested a modernity of process, and offered opportunities for both defence and attack, but also, with the requirement for the relevant infrastructure and equipment, constraining the options available.

Prior to the large-scale use of steamships, maritime supply had already been used to provide key logistical capability, as when the Russians advanced from the Danube to Adrianople in 1829, supply by sea freeing them from the risk of being cut off. So also with the French forces in Algeria from 1830. In contrast, inland forces, both in the age of rail and before it, faced far greater difficulties. The British force in Kabul, Afghanistan in 1841–2 was starved out, being forced into a retreat in which it was destroyed. Very differently, in 1878–9, Sir Frederick, later Lord, Roberts, was besieged outside Kabul, but he knew he had to stay, prepared a strong position, laid in food supplies for winter, and survived. Yet, the supply system was also a serious military encumbrance. Thus, in this conflict, the Second Anglo-Afghan War (1878–80), the British had to deploy 15,000 troops to guard the route between Peshawar and Kabul.[42]

The maritime dimension offered logistical capability as part of a key element of power-projection: the ready ability to lift to and from, an ability that was to be greatly enhanced when steamship transport increased volume and capability. There was a clear relationship as well with the enhanced prospects for the logistics of command in the shape of messages sent by rail, steamships, and the telegraph. The mobility of the last was improved by developing 'trains' with insulated wires and poles on wagons designed to extend fixed telegraph lines to advancing forces. Underwater telegraph cables provided strategic capability, especially for Britain prior to 1914, for example in the Anglo-French 'Fashoda' crisis of 1898 over competing African territorial claims.[43]

Logistics as an aspect of psychological mastery was also seen with the defeat of Paraguay by Argentina, Brazil and Uruguay in the War of the Triple Alliance from 1864 and 1870, a bitter struggle which devastated Paraguay, at the same time demonstrating the difficulty of fixing and destroying opponents. Thus, the invading armies found it hard to impose defeat on those of Paraguay, which was largely due to the political situation rather than the logistical flaws of the invaders.[44] Devastation as an aspect of warfare, as a 'counter-logistical' means, was the case more generally, and was an overlap with the politics of the broader struggle. In the 1870s, in the Cuban revolution against Spanish rule, Spanish forces employed harsh measures, including the relocation of the rural population so as to create free-fire zones and to prevent the rebels from gaining access to civilian aid, and also the killing of rebel families. In turn, the rebels waged economic warfare, destroying sugar mills and plantations.[45]

The destruction of crops and food stores was discussed in Charles Callwell's *Small Wars: A Tactical Textbook for Imperial Soldiers* (1896). He argued that tactics favoured the regular army whereas logistics favoured their opponents, and that, therefore, the need of the regulars was to fight not manoeuvre, and that their main problem was that of making and supplying the advance.

In the War of the Pacific (1879–82), Chile, once it had invaded Peru, used brutality to try to end opposition.[46] This conflict served as a reminder of the multifaceted character of logistics. There was the need by Chile to support its forces in the very difficult environment of the extremely arid Atacama Desert and, once Peru had been invaded, both in its lowlands and in the high-altitude Andes where moving supplies was far more difficult, for reasons of both terrain and less oxygen. There was the need to protect the logistical system from opposition by both regular forces and guerrillas, and there was the determination to hit the logistics of opponents. This situation overlapped with more general issues and methods of regular and counter-insurgency warfare, and serves to underline the difficulties of seeking to separate out logistics as a distinct question of needs and means focused simply on supplies. In particular, discussing supplies without their protection is a meaningless approach to the subject.

More generally, alongside 'long-term' logistical warfare in the form of attacking the food supplies of opponents, one also seen with American conduct toward Native Americans,[47] there was a widespread pattern, one in which supply problems played a role, in which rapid advances were the norm, while problems of supply encouraged the high-risk tactics of frontal assaults. In addition, there was no time to bring up heavy artillery in order to attack fortified positions. Just-can-manage supply systems proved to be an aspect of Latin American conflicts as well as those of Western imperial forces, notably in Africa. The politics of such logistics included the need to protect supply routes by means of local alliances, and, separately, to engage with difficult environments.

At the same time, improvements to supply systems could be of major value.[48] In Britain, the formation in 1888 of the Army Service Corps (part of a process of reorganisation begun in 1869), was of importance.[49] Improvements in logistics proved of particular significance when campaigning in the tropics: the invention of canned meat, dried milk powder, evaporated milk and margarine in the 1840s to 1860s, all highly significant due to the lack of refrigeration, changed the perishability and bulk of provision, making it easier to operate for longer periods between revictualling. Margarine derived in 1869 from a competition by Napoleon III to find a non-rancid substitute for butter for his army. Mechanical water distillation was also important, and helped overcome the major problem of drinking untreated local water, a problem that had had damaging consequences for Napoleon I's army in Russia in 1812. Medical and hygiene advances were also important to campaigning in the tropics.[50]

The warfare envisaged in Europe focused also on rapid advances in short conflicts,[51] but the scale of forces planned for and prepared was far greater. This, in turn, drove logistical requirements although again the scale was linked to the prospect of a quick war, like the Franco-Prussian (German) and Russo-Japanese ones of 1870–1 and 1904–5 respectively, and not the long conflict seen with the First World War (1914–18). The

German assumption was of front lines and civilian compliance, which helped focus German anger with opposition from *franc-tireurs* in their advance into France in 1870–1. So, even more, with the Austrian invasion of Bosnia in 1878, an invasion that suffered from guerrilla attacks on its supply convoys as they were moved forward on poor mountain roads. This made rear-area security a key issue.[52] That was also to be the case for the Germans in the Soviet Union and notably so from 1942 after opposition was encouraged by the Soviet winter offensive of 1941–2. Yet again, this serves as a reminder of the wider military and political dimensions of logistics. In terms of logistical strain, the factor of scale prior to the First World War was not simply that of numbers, which, in the case of the Russian army in 1900 was over one million strong excluding reservists. There were also the problems posed by qualitative changes, the most obvious the greater power of artillery which in part was made necessary by the enhanced strength of fortifications. Capacity was one issue, and an emphasis on speed in order to achieve or at least demonstrate victory not least to domestic audiences for whom cost issues remained of major significance. Costs were cut by speed.[53]

As guns and ammunition were heavy and could best be moved in the campaign zone on paved roads, advance by such roads became an adjunct to mobilisation by railway. The need to match plans to communications was therefore accentuated by the role of artillery. In the Russo-Japanese War, the Japanese benefited from organising an auxiliary service corps of reservists to provide logistical support and thus avoid the inadequate support provided by civilian contractors and labourers in the Sino-Japanese War of 1894–5. Having suffered from poor logistics in that conflict, the Japanese improved logistics and supply trains. Nevertheless, in 1904–5, there was also reliance on the traditional method of impressing large numbers of Chinese and Korean labourers. Despite its logistical efforts, Japan was challenged by having to campaign longer and further into Manchuria than had been envisaged, and was also unable to produce and distribute sufficient rations to prevent the loss of nearly 6,000 troops from beriberi as a result of vitamin B1 deficiency.[54]

Yet, alongside limitations, it was the logistical capabilities of the period that were more significant. The British ability in 1882 to move 10,000 troops and their support services to Egypt within forty days of the expedition being authorised was impressive, but, between 7 October

1899 and 30 January 1900, during the Boer War, 112,000 regulars were mobilised and sent to South Africa. In contrast, France had to rent shipping in order to invade Madagascar in 1895, and the United States faced major problems in mounting an invasion of nearby Cuba in 1898.[55] Similarly, enhanced capability expanded the possibilities of planning, with the Admiralty Staff in 1900 presenting Emperor Wilhelm II with Operations Plan III, the first German plan for war with the USA. By 1909, American battleships were being designed with larger coal bunkers allowing a steaming radius of 10,000 nautical miles, which was a major increase on the situation in the 1890s.[56] However, the American fleet's circumnavigation of the world in 1907–09 revealed serious shortages in infrastructure in the Pacific, as well as a lack of fleet colliers. To ensure supplies for his 1898 attack on the Spanish squadron in Manila Bay, Commodore George Dewey, the American commander, had purchased British merchant colliers, providing floating logistical support.

Change is a central theme in this chapter. There was a steady increase in logistical carrying capacity, speed and range. Change, however, did not eliminate problems, but, rather, generated new ones. Logistics thus remained difficult and, as ever, subject to the systemic problems, notably those of politics, tasking, and the military and environmental contexts. Alongside limitations, Western armies (and navies) do appear to have been more successful than hitherto and than most opponents, in freeing strategy and operations, not least in colonial campaigning, from the pursuit of supplies, an ability which fostered planning. Leaving aside the ethical question whether this amounts to 'superiority', as there is nothing inherently better about military success, Western military superiority on land, however, was more conditional and less secure than is generally implied. Moreover, the very size of the armies deployed increased the logistical burden, the combination encouraging, on a long-standing pattern, an emphasis on speed in deployment and action. Thus, in 1877, successive, costly attempts to storm Turkish-held Plevna in Bulgaria, in part reflected Russian logistical limitations. The First World War was a development of this situation. Critical to warfare for the period covered in this chapter, but also for 1914, was that strategic movement, by steamship and rail, was in many respects already at the standards of the mid-twentieth century, but that once the armies left the railheads they were still moving by boot leather and horseback.

Chapter 7

The First World War, 1914–18

The apex of logistical activity is generally seen as the world wars, the major subject in this chapter and in chapter nine. Foremost, this view arises because of the scale and range of military activity, which were stupendous, and notably so in the Second World War. Yet, there were also other factors leading to this emphasis. The two world wars, particularly the second, were seen, at the time and subsequently, as the encapsulation of total and industrial warfare. The second saw the mobilisation of American industry and a major American commitment that helped mould the geopolitics of the succeeding decades. American publishers, writers and the American book market proved particularly influential in military history, and, with reason, saw the American logistical system in its widest sense as in part instrumental to success and as defining logistical capability, with, as it were, 'Rosie the Riveter' on the Home Front winning the war. It would be naïve to underplay this logistical strength, but it should be located in a broader assessment of the reasons for Allied victory.

And so also for the First World War, in which logistics was significant from the outset, notably in the capacity of respective railway networks to support the mobilisation of troops and their supplies. Logistical requirements in the war of manoeuvre in 1914, however, were different to those when the front lines became more fixed with trench systems, although manoeuvre-tempo logistics continued to be significant as with the Central Powers'[1] conquests of Serbia (1915) and Romania (1916).

After the initial attacks in 1914, with their use of prepared stores in a conflict that was expected to be bloody but relatively short, logistics was a matter not only of campaigning but of overall production capability. Logistical links were important not only within individual states, but also in getting alliance systems to cooperate. Thus, they were crucial in securing the *matériel* on which Allied war-making depended, and, indeed, were in the front line of both wars with German naval activity increasingly focused on attacking the Atlantic shipping routes that were crucial to the allies. Neither Britain nor France had an industrial system to match that of Germany which, by 1914, had forged ahead of Britain in iron and steel production. Indeed, the Allies were dependent on America for machine tools, mass production plants and much else, including the parts of shells. American industrial output was equivalent to that of the whole of Europe by 1914, and the British ability to keep Atlantic sea lanes open ensured that America made a vital contribution to the Allied war effort before its entry into the war in 1917.

This contribution underlined the significance to logistical systems of protection by means of convoying, one seen in many contexts across time, and on land as well as at sea.[2] Transoceanic trade and naval dominance, with the oceans as their interior lines, allowed the British and French to draw on the resources not only of the United States, but also of their far-flung colonial empires. Logistics reached to every farm, forge and forest and was understood in that light, not least with the direction of production by government.[3] Britain's logistical system was also that of a country that could not feed itself; nearly two-thirds of British food consumption was imported. As a result, German U-boat (submarine) warfare was a form of counter-logistics.

With the imposition of blockade, which was a naval and an economic process, logistical warfare was also declared on Germany (and its allies), and its access to trade was also gradually reduced as the Allies steadily increased pressure on neutral powers to stop the lucrative practice of re-exporting their imports to Germany, particularly on the Dutch not to re-export American food landed in Rotterdam. This economic warfare was supported by a system of pre-emptive purchasing, for example of Norwegian fish and pyrites, that was important to the international control of raw materials, as well as greatly influencing neutral economies. In particular, cutting off trade with Germany lessened American economic and financial interest in its success.

The Germans, instead, pursued a more primitive logistical system, that of seizing the resources of conquered areas. This was notably the case with the coal and iron of Belgium and north-east France in 1914, the grain and oil of Romania in 1916, and grain and other resources in Ukraine in 1918. However, advancing to seize these resources in 1918 was a factor in the delay by the Germans in transferring forces from the Eastern Front to the

Western that year, a delay that was harmful to German capability on the latter. Moreover, in 1918, the collapse of Bulgaria, and the consequent Allied move northward from Salonica (Thessaloniki) cut Germany off from the Romanian oilfields.

In turn, for all powers, protecting the logistical system and, more specifically, the increase in armaments production, with the British Royal Arsenal at Woolwich employing over 50,000 people by the end of the war, were necessary because of the unexpected length of the conflict and the extent to which artillery in particular became more significant. At sea, logistical protection involved a range of naval tasks, both defensive – notably convoy protection and minefields – and offensive – in seeking to attack German surface raiders, submarines, and submarine bases, including by the laying of mines, a weapon that developed considerably from 1884.[4] Munitions factories were protected from bombing by anti-aircraft guns.

The stabilisation of front lines in the form of trenches required heavier, high trajectory artillery supplied with plentiful shells. Indeed, artillery was the real killer of that war: estimates suggest that high explosive fired by artillery and trench mortars caused up to 60 per cent of all casualties, and the supply of both guns and of ammunition for them therefore became key issues. J.F.C. ('Boney') Fuller, later a prominent military thinker, but then a British officer on the Western Front, wrote to his mother in August 1915, 'One salient fact stands out throughout history … whichever side can throw the greatest number of projectiles against the other is the side which has the greatest chance of winning'.[5] A month later, the British journalist Philip Gibbs linked the struggle to produce goods on the 'Home Front' to the major (but in the event unsuccessful) British assault on German lines at Loos on the Western Front:

> All the batteries from the Yser to the Somme seemed to fire together, as though at some signal in the heavens, in one great salvo. The earth and the air shook with it in a great trembling, which never ceased for a single minute during many hours. A vast tumult of explosive force pounded through the night with sledge-hammer strokes, thundering through the deeper monotone of the continual reverberation…. This was the work of all those thousands of men in the factories at home who have been toiling through the months at furnace and forge.

They had sent us guns, and there seemed to be shells enough to blast the enemy out of his trenches.[6]

In response to the development of what was an existing need and system, artillery remained a centre of logistical attention. The scale was unprecedented. Thirty-seven million shells were fired by the French and Germans in their ten-month attritional contest for the fortress-city of Verdun in 1916. The Italians, not the most industrialised of powers and somewhat short of artillery, nevertheless deployed 1,200 guns for their attack on the Austrians in the Third Battle of the Isonzo in October 1915, the British 2,879 (one for every nine yards of front), for theirs near Arras in April 1917 and the Canadians 1,130 the same month at Vimy Ridge, also on the Western Front. The German 150 mm field howitzers could fire five rounds per minute.

Such a use of artillery ensured that the cost of offensives soared, while, in 1915, as a result of a failure to anticipate the problem, the availability of shells became a political issue in Britain, which lacked an adequate munitions industry, and a serious problem in France (which had such an industry) and Russia. The need for shells for heavy, rather than field, artillery, and high explosive as well as shrapnel shells, exacerbated the issue. To deal with these and related problems, large sections of economies were placed under governmental control and regulated in a fashion held to characterise military organisation. For example, the Ministry of Munitions, created in Britain in 1915, was as much part of the military organisation and logistical chain, and as vital, as the artillery it served. David Lloyd George left the Treasury, a key ministry, to become the first Minister of Munitions and made his name as a wartime leader in this role. Responding to the need to produce more artillery shells, Lloyd George bypassed established procedures by enlisting entrepreneurs in the cause of production. Moreover, a political purpose was served as Lloyd George used his ministry to demonstrate his belief that capital and labour could combine to patriotic purpose.

So also with the French Ministry of Munitions under, first, Albert Thomas and, from September 1917, Louis Loucheur. The French government also controlled bread prices, while state-supervised consort directed the allocation of supplies in crucial industries, although the production of munitions was left to entrepreneurs. State control was widely extended in France, including over the chemical industry and the production of footwear, the latter necessary for soldiers' boots.

There were also important qualitative as well as quantitative improvements in the production of munitions in Britain as a result of involving the trade and introducing co-operative group manufacture, whereby each manufacturer within the group made some of the components of the munition. This allowed inspection to be carried out at one location, the premises where the components were put together, instead of at the premises of each manufacturer, thereby speeding up production. The increased skill of the trade reduced rejections, and the inspectors from the Outside Engineering Branch of the Ministry of Munitions ensured that production increased to meet demand. In 1916–17, various unnecessary components in British shells were eliminated, which also speeded up production. One of the problems faced by the British in 1915–16 was the poor quality of the shells. So great was the improvement brought about by better working practices, greater experience, and the inspectors, that prematures fell to 0.004 per cent, one in 250,000, the best rate in the Allied armies, and a formidable achievement of industrial application. The quantity of weaponry available to the British rose greatly in the last quarter of 1917. Moreover, a greater application of ranging systems helped lead to the more effective use of the artillery available.[7]

With the war lasting far longer than most had anticipated or planned for, improvements in manufacturing capability had to be organised, financed and translated into necessary resources for the logistical chain. For example, British fire-support for the infantry improved greatly, thanks to the use of the three-inch Stokes light infantry mortar and the Lewis man-portable machine-gun (about 30 per battalion in 1918), and the development of reliable hand grenades (Mills bombs) which could also be fired as rifle grenades, as well as the development of relevant tactics. Production rose rapidly to meet demand. In the last quarter of 1914, only 2,164 hand and rifle grenades were produced in Britain, and, although the figure had risen to 65,315 in the first six months of 1915, it was still well below demand. Only in October 1915 did the output of the Mills No. 5 grenade meet demand when it passed 300,000 a week. Already in 1915, the monthly demand for percussion grenades alone had risen to 252,000. British output of mortar ammunition rose from 50,000 rounds in April-June 1915 to 2,185,346 million rounds for April-June

1916. A total of 11,052,451 grenades were delivered from Britain in the second half of 1916, and that year the British output of hand grenades was close to 29 million.[8] In addition, deliveries of trench mortars from Britain rose, from twelve in the last quarter of 1914 to 2,145 in the last quarter of 1917.

These were large quantities but the relatively fixed nature of front lines eased the pressure on logistical systems, making it easier to provide a fixed destination for supplies. However, the distribution of supplies to front-line units varied, with Austria, Italy, and Russia, for example, less successful than wealthier Britain and France. Networks of light railways were created to aid logistics near the front line. Rail systems, however, were highly vulnerable to shelling.

Alongside organisational and resource capabilities, logistical systems relied on applied knowledge, from unit needs to overall calorific and nutritional requirements, and methods of food preservation. Energy values, not taste, were to the fore, ensuring that the British relied on bully beef and biscuit.[9] John Monash, an Australian brigade commander and before then an engineer, wrote to his wife Hannah from Gallipoli in 1915, describing the extent to which troops were part of an industrial process, with a specialisation of function and intensity of organisation that matched, and connected with, features of contemporary industrial society:

> We have got our battle procedure now thoroughly well organised. To a stranger it would probably look like a disturbed anti-heap with everybody running a different way, but the thing is really a triumph of organisation. There are orderlies carrying messages, staff officers with orders, lines of ammunition carriers, water carriers, bomb carriers, stretcher bearers, burial parties, sandbag parties, periscope hands, pioneers, quartermaster's parties, and reinforcing troops, running about all over the place, apparently in confusion but yet everything works as smoothly as on a peace parade, although the air is thick with clamour and bullets and bursting shells, and bombs and flares.[10]

These troops required a range and quantity of equipment that could not be obtained locally: from timber for the construction of trenches to spectacles.[11] The absence of plastic ensured that packing was usually in timber crates which added to the weight that had to be moved, and, therefore, the energy demands and transfer commitment on the transport system, in the shape of the ease of loading, unloading, and transhipment. The focus in the discussion of logistics was to be on the supply of artillery shells, but food remained a key element. On a pattern as old as siegecraft, this ensured that the sustainment capabilities and supply links of fortresses were a major factor, as with Austrian-held Przemyśl in the face of Russian attack in 1914–15: its food stocks proved deficient.[12] Another aspect of continuity was that the British Army still shipped to the Western Front a greater weight of horse fodder (largely for transport horses) than artillery shells.

At the same time, there were to be totally new supply networks, as for air power, with, in this case also, a rapid increase in the scale of operations. In addition to airframes, engines and fuel, there was a range of requirements, notably ammunition and cameras.[13] This was for all practical purposes the development of an entirely new major weapons system from scratch in four years, embracing industry and mechanics.

It was different from the tank in the sense that a motor industry already existed in 1914. Nevertheless, tanks posed new logistical requirements. A main theme in the discussion of tank warfare is the limitations imposed by logistics when set against the ambitions of the protagonists of tanks such as J.F.C. Fuller.[14] Thus, scale was not the sole issue, as there were also new challenges.

As in most coalition wars, logistics proved most effective when international co-operation was in active play. Thus, French heavy industry, including the production of munitions, owed much to the supply of coal, iron and steel from Britain, and shipped in British ships; and the British introduced convoys first in February 1917, to protect ships carrying coal to France, doing so in response to calls from the French government. In part, these supplies compensated for the loss of French production capability to German occupation. At the same time, the British and French armies were supported by their respective economies, including the timber shipped across the Channel for the British trenches.

The provision of coal, absolutely crucial for manufacturing and railways, was an important instance of an Allied co-operation seen also, in 1916, with the establishment of the Inter-Allied Bureaux of Munitions and Statistics, which were designed to help co-ordinate and plan munitions provision, including purchase in America. These bodies led, at the start of 1918, to the Inter-Allied Munitions Council, although, despite the Inter-Allied Tank Committee, tank production proved an example of limited co-operation. On a global scale, the Allies directed most of the world's shipping, trade and troop flows. The Allied Maritime Transport Council oversaw an impressive system of international co-operation at sea, allocating shipping resources so that they could be employed most efficiently, which was important economically and also in lessening targets for German U-boats (submarines). The Wheat Executive provided another aspect of this international planning. Far more than planning was involved. For example, American and British technicians and locomotives helped to sustain the French rail network, while the shortage of French deep-water harbour capacity ensured that many American troops landed in Britain before transhipping to smaller ships for crossing to France.

When America sent its army to France, the pressing need for troops and the strains of developing the requisite manufacturing and shipping capacity, ensured that American forces reached Europe without the necessary equipment. The French and British equipped the Americans with French and British artillery and other munitions: the Americans were particularly dependent on the French for artillery, machine-guns and food, and on the French and British for tanks and training in their use. The Americans had their own design of hand grenade, but used the British Brodie helmet, not one of their own design. The American infantry were saddled with the inadequate Chauchat light machine gun, but, more helpfully, the Americans also had the French 75-mm gun and the VB rifle grenade. Most of the artillery, tanks and aircraft used by the Americans in the Meuse-Argonne offensive were provided by the French,[15] who produced more munitions than Britain in 1918, and, as a result, were able to fire formidable numbers of shells that year, both in opposing the German offensives and, subsequently, in support of their own attacks.

War brought regulatory change in America to forward logistics, with the rail network brought under government control in 1917. Mobility took different forms, but throughout was designed to improve movement and output for the labour used. Thus, America sent tractors and tractor ploughs in numbers to Britain in 1918, and, although their introduction involved problems, they helped deal with the manpower difficulties in agriculture caused by military recruitment and also by work in construction and munitions. A different use of motor vehicles was provided by lorries (trucks) which produced mobility for both troops and supplies, and underlined the value of oil supplies. Lorries were important both to the logistical system as a whole and in so far as particular operations were concerned. In 1916, the French used lorries on the *Voie Sacrée* to supply Verdun and the Italians used lorries and railways to bring up troops and supplies to hold back the Austrians advancing from the Trentino. By 1918, the French army had nearly 90,000 motor vehicles, while the Germans had only 40,000.

The 72 kilometre (45 mile) *Voie Sacré* from Bar-le-Duc to Verdun was particularly necessary because all the standard-gauge rail links to Verdun were affected by earlier German advances. As a result, the road was widened in 1915 so as to be able to take two lines of lorry traffic moving in opposite directions. In 1916, all troop movements on foot and horse drawn traffic was banned, so that the road could build up to a four-vehicles-a-minute capacity. This posed formidable management and maintenance challenges, the latter handled by labour battlalions and vehicle servicing, including tyre renewal and breakdown lorries. 3,500 lorries and 800 ambulances were in use on the road, which, alongside a narrow-gauge single-track railway that was upgraded, supported nearly half a million troops at Verdun. Bar-le-Duc had good supply links, including by canal.

The significance of transport links helped increase interest in bombing as a means to ensure the disruption of those of opponents. Thus, in 1918, in support of Britain's advance northward in the Near East, British air superiority prevented the interception of British bombers which were used to bomb rail junctions and telephone exchanges in order to disrupt communications, and to destroy the Turkish forces when they retreated.

Unlike in the Second World War with its emphasis on strategic air assault in order to destroy the opposing economy, a process beginning with the failed German attempt to do so against British ports in 1940, naval blockade was the key means in this war to the end of economic warfare; although in part the damage that could be attributed to the Allied blockade was an aspect of the more familiar issue of prioritisation. Thus, for example, the logistical capability of the German system was affected by the determination to raise army manpower and armaments production. Due to a lack of miners, coal output fell in 1917, and both this and the lack of workers affected the rail system which, from 1916, was also under great pressure due to the focus of steel production on armaments. The cumulative impact of such shortages was a run-down in the German economy and its growing atomization, which hindered attempts to co-ordinate and direct production.

In response to German submarine warfare, Britain moved toward a total war mobilisation of the resources of society, and, as part of the logistical infrastructure, Britain sought to shorten the supply chain by manufacturing or growing imported goods at home. With the Corn Production Act of 1917, the government created a Food Production Department and imposed a policy of increasing the amount of land that was ploughed, replacing meat production by cereal. Coal and food were rationed. The results of focused production were seen for Britain in the front line, both in resources and in a marked improvement in logistics.[16]

Across Europe, many sections of society were not brought under the direction to which munitions production was subject, but can still be seen as part of the informal organisation of militarised states, with governments extending their regulatory powers in order to ensure that resources were devoted to war and to increase economic effectiveness. Nevertheless, in an example instructive for previous societies, the Russian economy, in contrast to that of Germany, was poorly managed and wasteful, with industrial production and transport both in grave difficulties by 1915. A serious munitions crisis affected Russia that year, and the Russians depended in part on French supplies. There were efforts by Russia to improve organisation, but they could not rise to the burden of the war which was exacerbated by defeat and the resulting loss of territory and resources. The supply of food and ammunition to units at the front was frequently insufficient, and if the same was true of Austria, the Germans in contrast were better prepared on the Eastern Front.[17]

That posed a logistical challenge greater than the Western Front, not only due to its longer extent but also because of weaker communication networks, fewer local resources, and a harsher climate. This situation was exacerbated in the ancillary campaigning zone in the Balkans, as with the logistical requirements for the Central Powers' invasions of Serbia (1915) and Romania (1916), in the latter of which it was necessary both to advance across the Danube and through the Carpathians.[18] So also with Russian successes against the Turks in the Caucasus.

With its far more developed industry, Russia itself was able to produce a range of resources that was greater than that of Turkey. Although the Turkish army at Gallipoli had enough supplies, especially, despite claims to the contrary, ammunition, Turkey was hit hard by the inability to maintain the rural economy in the face of the demands for recruits, animals and food. Military control over logistics extended to provisioning, but this control could not deliver results other than short-term.[19] Yet, Turkey was able to fight on for longer than Russia. This contrast reflected a number of factors, notably political cohesion and resilience, but the extent to which the well-resourced British forces were only able to bring significant force successfully to bear in 1917 and then in essentially marginal areas of the Turkish empire – southern Palestine and central Iraq – was important.

Outside Europe, the logistical situation during the war could be very different to that of Europe, not least because of swifter movement across longer distances. Thus, light lorries were used in 1916 in the British conquest of Darfur whose Sultan had heeded Turkish calls for Islamic action. However, in sub-Saharan Africa, where there was an attempt to engage in Western technological warfare in areas with minimal infrastructure, the British were heavily reliant on human porters, and these had a high death rate. Over a million men served the various powers as carriers and impressed workers in Africa. In the long Allied campaign to conquer German East Africa (Tanzania), the Germans benefited from invading the Portuguese colony of Mozambique in 1917 and seizing resources there. Their victory at the battle of Ngomano on 25 November produced prisoners who were used as porters, and the capture with them of uniforms, rifles and ammunition, with which the force was now re-equipped.[20] It then campaigned in Mozambique for several months.

The last year of the war saw logistical systems coming to full capacity, but some under acute pressure. The rise in German shell production in 1917, in response to the British use of heavy artillery during the Somme offensive in 1916, was important in providing the material for the 'iron hurricane … the avalanche of missiles'[21] that launched the Aisne offensive on 27 May 1918 and was preceded by a heavy bombardment of two million shells from 6,000 guns in four and a half hours. However, advancing German troops were affected by logistical problems, outrunning their supplies, by the failure of artillery and machine guns to maintain the same rate of advance, by the German inability to capture the key railway terminals, and by the systemic extent to which the Germans had damaged their own production and factory-to-front supply chain. The greater effectiveness of British and French logistics, in turn, was seen in the 1918 Allied offensives, offensives that reflected the Allies ability to develop the handling necessary for offensives carried out by very large forces taking part in theatre-wide campaigns and with the combined-arms methods necessary to break through the defensive zones the Germans had developed.[22] By this stage, the Allies were making much greater logistical use of lorries, which offered capabilities that mules and trains lacked, notably speed and range, and did not require fodder or coal, (but, however, they needed oil, maintenance and roads).

Marshal Foch's new attack methods deliberately dislocated the German immediate military supply chain at the operational level,[23] and this, plus desertion, left only a handful of very understrength German divisions as still reliable. The British were meanwhile fighting a 'rich man's war', which the Germans argued was a *Industrieschlacht*, in which they claimed their fighting ability was out-resourced by their opponents; but British production exceeded their transport capability which was affected by the scale of demand and the friction of campaigning. In September–November 1918, they were less and less using their modern technology, notably heavy artillery, and reverting to field guns. In part, as also with tanks, this was due to them outrunning both their transport and their ability to keep their technology moving, while guns and rifles were good enough to defeat the very weak Germans.

The remarkable parallel with the Allied advance in France in August–October 1944 is significant as men who were majors in 1917 were generals in 1944, and thought they saw history repeating itself. The British army in November 1918 would almost certainly have needed a logistical pause; and anyway the weather was getting much worse. Foch and Haig both understood the need for such a stop. Although the Allies were coping reasonably well with the logistic implications of the increased mobility caused by German weakness, this mobility also posed major challenges. Movement forward on this unprecedented scale was a major logistical strain.[24]

As with J.F.C. Fuller's bold plans for the use of tanks in 1919, plans that underestimated the issue of maintaining fuel supplies, it is unclear how well the logistical system would have coped had there been a full-scale advance into Germany, not least with the Germans continuing to destroy communication links, a policy that necessitated the large-scale Allied use of Labour Battalions, notably to repair railways. The German alliance system was collapsing in late 1918, but there was no equivalent to the commitment posed by the Soviet advance in 1944–5. If, therefore, the war had continued, the logistical support required by the Allies would have been to overcome a still resisting opponent and with major physical obstacles to confront, such as a Rhine crossing. Room for optimism on the logistical dimension is therefore mixed. An incremental practice would have worked, but the scale of the advance – from the Swiss to Dutch frontiers, and with an attacking army expanded by the far greater size of American units – would have been formidable whatever the challenge.

Chapter 8

The Interwar Period, 1918–39

The war was followed by a very different period in military history, with civil conflict and imperial control to the fore from 1919. The emphasis was on activity (albeit small scale compared with the world war), the seizure of key political centres, and raids, rather than on sieges or on staging battles from prepared positions, these reflecting a stress on the offensive, not least in order to seize resources. In the case of civil wars, such as that in Russia, logistics were very improvised, and therefore frequently brutal. Even, very differently, when violence was avoided in civil disputes, the provision of supplies was still significant. Thus, in Britain during the General Strike of 1926, troops and police drove buses and underground trains, kept power stations in operation, worked on the docks, and distributed food.

In the formation of new states after the First World War, it was necessary to establish new armies, and therefore logistical organisations. As with the Americas with the crisis and fall of European empires in 1775–1824, this process entailed disputed legitimacy in the case of authority, as well as contested power in that of establishing new structures of government and then employing them for militia purpose. The mobilisation of new forces ensured tensions in society including at the very local level, with protection a key element of military activity.[1]

Armies that had taken a role in the First World War, as well as neutrals then, faced the challenge of imperial warfare in difficult environments. In 1921, both Greek and Spanish forces, advancing respectively into the interior of Turkey and Morocco, were greatly affected by supply problems, notably of water and food, which accentuated disease and exhaustion. These shortages were exacerbated by attacks by opposing irregulars on supply lines, while the Turks and Moors proved better able to support themselves, as a British military observer in Turkey in the winter of 1919–20 noted.[2]

In Morocco in 1921, the Spanish force ran low on ammunition, which the rapid-firing rate of modern firearms made more serious. Defeat at Annual that summer also saw a major transfer in supplies, with the Moors capturing at least 14,000 rifles and at least 100 pieces of artillery, as well as the 1,500 mules crucial to Spanish logistics. More generally, the Moors were not dependent on the cumbersome supply routes of their opponents, while Spanish positions were supplied by convoys that were easy to harry. A correspondent of the *Times* returned from Morocco, and noted in the issue of 20 January 1925, in an article more generally relevant to the conflicts and logistics of the period, both imperial and elsewhere:

> Four years of warfare in European trenches adjusted the modern mind to the idea of a united front, along which all who fought on the same side could join hands and muster in endless lines without a thought for the safety of their communications and only friends or vanquished behind them.
>
> In order to understand the present situation in Morocco and the events which have led up to it, it is first of all necessary to blot out this picture of an infinite trench and replace it by that of a front – if such a word is permissible in this case – consisting of a series of islets scattered over a country which may have been conquered, but has not yet been pacified. Some of these islets were known as base camps or positions, others were called blockhouses ... the Moors were at liberty to select both the target and the moment for attacking it.[3]

This was to be more generally the logistical pattern, but was one that was interrupted in the world wars when more usually (although not always) there were continuous front lines. The absence of continuous front lines was a feature of the civil wars in China, Spain, Nicaragua and elsewhere, and there were frequently advances without any or well-developed flanks, a situation seen more recently in, for example, conflict in Congo from the late 1990s. The logistical strain varied depending in part on the size of forces that had to be supported and the nature of the physical and built environment. Thus, in China in the 1920s, the national rail system existed as a steel framework connecting the new, industrial, cities and acted as a kind of modern battlefield: most of the fighting was along, or around, in, or for, it. Thus, in 1924, We Peifu, the head of the Zhili clique of warlords, depended on the rail system in order to concentrate his forces against Zhang Zuolin, the Manchurian warlord. At the same time, this activity posed a serious logistical strain. *The Times* noted on 26 September 1924:

> Nearly 300 train loads of troops and supplies have gone forward in 20 days, a considerable performance for a single-line railway. This result has been made possible by withdrawing practically the whole of the rolling stock from all the railways north of the Yangzi river, thereby completely stopping goods traffic and permitting only a precarious minimum mail and passenger traffic.
>
> There is much confusion at the concentration point and the empty trains are not returning. Military officers are seriously interfering with the railway administration and enforcing their demands at the point of the pistol. In one case where a number of officers were disputing for precedence the difficulty was solved by joining five trains together and sending them forward as one train, measuring one and a half miles long. Speed was reduced to a walk, but the 'multiple tandem' arrived safely to the satisfaction of the 4,000 soldier passengers. The locomotive crews having objected to long hours are working double shifts, with sentries to see that the men off duty sleep soundly and do not gallivant in the darkness.[4]

The importance of the railways helped ensure that cutting them became a purpose of military moves, while advances were correctly described as moving along the railway, such as the Longhai Railway along which the Central Army marched in 1930 in the Central Plains War. This process was encouraged, as also in Russia and elsewhere, including Estonia, Lithuania and Poland, by the use of armoured trains, which, developed from the 1860s, provided a way to deliver firepower and also to provide protection. Railway junctions proved key operational goals and garrison points, in China, and notably so if the junctions were also cities as the latter produced revenues and offered legitimacy.

The significance of the railways was enhanced not least by the contrast with the very different communication system used in their absence – coolies, or human porters, on in which compulsion played a key role. The *Times'* Shanghai correspondent reported:

There is great indignation swing to the merciless kidnapping by soldiers of thousands of peaceable Chinese for transport and trench-digging in the districts about Shanghai. Male inhabitants of the surrounding district have flocked in a state of terror into the settlement to escape the press gang. Similar reports come from Nanking and Soochow. Both sides are equally blameworthy of this disgraceful oppression.[5]

Men were also seized to dig trenches and build earthworks, a key infrastructural requirement that is now largely provided by mechanical equipment. The impact of the season, and notably rainfall, on campaigning was another of the traditional aspects of Chinese logistics, although it was also to play a role in modern campaigning. Most troops in China were infantry, which underlined the significance of railways for transport and logistics, although the significance of mobility and the paucity of modern road vehicles helped ensure the continuing role of cavalry.

There was also an overlap in China with conflict at the local level which brought in a different type of military activity with the associated logistics. Large bandit gangs were locally significant. They lacked the ability of the warlords to use resources and loans in order to obtain weapons from abroad, an ability grounded on their territorial bases, but challenged the security of the territorial positions from which the warlords raised these resources of men, money and food. As an overlap between warlords and bandits, both oppressed the population; this overlap meant that bandit forces could be absorbed into the warlord forces while particularly undisciplined warlord units could become bandits. Warlords concentrated on cities and towns, to control their wealth and benefit from their importance for communications.

The 'progressive' narrative of logistics, the emphasis on development through change can draw on the increase of scale in Chinese armies in the 1920s, the purchase of modern weapons from Europe, and the use of foreign experts, but it is likely that the transformative character of both arms and advisers in China has been exaggerated, and notably so with the warlord forces. At all levels, logistics were very important because troops were motivated by 'the rice-banks' at a time when soldiering guaranteed employment, local bonds were highly significant, and armies were often poorly integrated into society.[6] In practice, looting and extortion were key means of raising supplies. In the early 1930s, the logistical nature of civil war in China changed because the Communists were based in highland areas and not in the lowland cities on which the warlords of the 1920s had focused. Mao Zedong, the Communist leader, argued that one of the keys to victory was partly relying on the enemy logistical chain, by everything from corruption and subversion to armed seizure through force.

Across the world, the 1920s and 1930s brought to the fore the question for armies that had always affected navies, namely that of the problems entailed in the supply and maintenance of machines, and also how to project and sustain industrialised warfare into regions that lacked the relevant infrastructure. Machinery cost much money, not only to develop, but also to introduce, supply and maintain. This was true both for complex machines, such as the aircraft used by the Americans to airdrop supplies to their Marines in Nicaragua, and also for more simple equipment, such as improved firearms. On the other hand, there were heavy costs entailed in providing for large numbers of troops, notably in training, paying, housing, clothing, and, particularly so in wartime, feeding and supplying them. Machines were seen as a way to deliver increased capability at lower costs, but that had major implications for supply systems; although these were not a simple product of the nature of the machine. Instead, doctrinal ideas about best practice were an important variable, as with consideration of how best to use tanks.

Discussions were, in turn, affected by the presentation of the recent war and consideration of the next one. This was clear with the assessment of tanks, but there was also, as a result of the First World War, a greater commitment to lorries/trucks as a part of the supply system, and notably in the absence of rail links. In 1926, Colonel George Lindsay, the Inspector of the [British] Royal Tank Corps, emphasised the need for mechanisation:

All civil evolution is towards the elimination of manpower and animal power, and the substitution of mechanical power ... in the army we must substitute machine and weapon power for man and animal power in every possible way, and that to do this we must carefully watch, and where necessary foster, those trends of civil evolution that will help us to this end.

Lindsay was certain that this capacity was linked to industrial capability: 'We are the nation above all others who can develop the mobile mechanical and weapon-power army, for we have long service soldiers and a vast industrial organisation'. Yet, road vehicles also faced problems across much of the world. Warning against British intervention in Turkish Armenia in 1920, the General Staff commented: 'before operations could commence, the Trebizond-Erzerum road, a major route, would require to be remade to take mechanical transport, an immense undertaking considering the distances and the grades'.[7] Separately, there were also issues of scale, Archibald Wavell, then a British brigadier, for example, doubting in 1930 the availability of sufficient aircraft to fulfil large-scale supply needs.[8]

By accentuating logistical problems, largely due to the pressures of keeping up with advances and supplying fuel, greater mobility had a major impact on planning needs and doctrine. The mobility was both the consequence of logistical achievement in the shape of the delivery of fuel and of keeping up with advances but also the cause of logistical requirements. This mobility also led to the development of an operational dimension to warfare, in part as a combined arms approach addressing the possibilities of the new weaponry.

Combined with the need for oil for aircraft and warships, this situation encouraged concern about the availability of sufficient oil, which, indeed, became a key element in logistics as well as operational planning. For some states, there was interest in the production of synthetic substitutes for imports. Germany's Four-Year-Plan, introduced in 1936, developed production of synthetic oil, rubber and textiles, while Japan stepped up synthetic oil production in 1938. This reflected the dominance of oil production by America, Britain, the Netherlands, and the Soviet Union. The global network of bases necessary for the coaling of steamships became less necessary when oil replaced coal as oil could be transferred from oilers to warships at sea, as with German surface raiders in the Atlantic, notably the *Graf Spee*. Nevertheless, bases continued to be significant, and were seen as such in American debates about how best to fight Japan and, in particular, what could be achieved without forward bases.[9]

Fuel was not the only issue. The General Board of the Navy also set out to enable repairs close to the fleet and did so by developing floating dry docks, notably the Advanced Base Sectional Dock-1. Carrying aircraft in carriers increased the range of spares required for routine maintenance requirements, and the forecast defect rates. Ships also deployed with fuels and oils and armaments relevant to the aircraft embarked, a situation amended as experience and situations dictated. To do so, logistics personnel had continually to look ahead as well as back. Lessons learned had to be applied to the future, and the key being to adapt the lessons to the new situation. In thus providing the most effective and efficient support possible within any constraints, not least of which was the financial element, they also had to consider scenarios and their possible impact. That was very difficult in the context of the 1920s and 1930s as there was no experience of fighting a carrier war. As well as carriers at sea, theorising about air warfare far outstripped the logistical capability for sustaining aircraft.[10] The same was true of theorising, especially by J.F.C. Fuller, about land warfare with tanks, conceived of as land fleets.

Logistical questions played a significant role in war-planning and, very differently, military exercises. There was recognition after the recent world war of the need for peacetime military-industrial planning for production to meet wartime expansion. In America, as a result of the National Defense Act of 1920, which had required the army to plan for wartime mobilisation, the Army Industrial College was established in 1924 to ease mobilisation. Then a Major, Dwight Eisenhower graduated from it in 1933 and subsequently served on the faculty. In 1946, the college became the Industrial College of the Armed Forces. Eisenhower had already contributed much to the 1930 Industrial Mobilisation Plan, supporting wartime price controls as a means of working of what was to be termed the military-industrial complex. The coordination of army and navy procurement was another cause he backed, although the necessary planning did not materialise in the early stages of the Second World War, with resulting chaos.[11] In contrast, the Japanese military followed a very different logistical policy, notably regarding the conquest of Manchuria in 1931–2 as a source of resources and industrial capacity.[12]

A contrast in methods of warfare and the related logistics were seen in China in the 1930s, for there were two different sets of conflict, the first the struggle between Guomindang (Nationalist government) and the Communists, and the latter between China and Japan. The Guomindang used the counter-logistics of scorched-earth, blockade, and the forcible movement of peasants to attack Communist-held areas. The Japanese showed from 1937 that it was not necessary to introduce mass mechanisation in order to conquer large tracts of territory. Lacking raw materials and industrial capability (which helped explain the strategic importance of gaining the resources of Manchuria – fighting to secure the resources they needed to be a power strong enough for its goals), Japan was technologically behind the Western powers in many aspects of military innovation, such as the use of tanks and motorised transport, and there were longstanding problems with Japanese army logistics. The shortage of lorries, a shortage that remained a problem for Japan throughout the war, made the army reliant on animal transport across most of China, as much of it was not accessible to river or railway. The shortage, as well as the bad state of the roads, ensured a heavy reliance on railways where they were present, and, as in 1940, these railways were vulnerable to guerrilla attack.[13]

The war in China from 1937, but to a degree even from the Shanghai Incident (battle) of 1932, challenged the Japanese stockpiles almost immediately. After general mobilisation in August 1937 as troops poured into the mobilisation centres of their regiments and units were dispatched to China, the vastness of the theatre of conflict there was challenging, and the heavy fighting in Shanghai that year depleted the resources of the Japanese expeditionary forces. Following the Japanese breakout and near encirclement of the Chinese forces and the difficult advance to Nanjing, many Japanese were ordered to live off the land. The pattern of operations in Central China was established, that of living off the land, and the routine impressment of Chinese labour to carry supplies when Japanese pack animals in the previously well-organised supply unit died. Many of the operations for the next couple of years were those of leaving a base city that was reasonably well-supplied and then extending operations beyond what they could carry so that they depended on the civilian population (or what they had left behind) to sustain operations.

The Chinese swiftly saw scorched earth policies and poisoning wells as ways to defeat such thrusts. Japanese logistical deficiencies thus included not only the limited resources that Japan could deploy but also, more specifically, the problems of transporting sufficient food as Chinese forces cleared the countryside of supplies to prevent the Japanese living off the land. In November 1938, the Japanese advance toward Changsha in Hunan was stopped not only by effective Chinese resistance but also by supply problems that contributed to malnutrition. The logistical crisis for

Japanese forces in China included a shortage of ammunition, as well as an inability to bring up supporting artillery. Soldiers were expected to carry nearly 30 kilos of equipment and rations. Changsha was not captured until 1944.

Logistical capability is in part a matter of the ability to act in response to constraints. The Chinese also lacked adequate motorised transport, but this was less serious because they were largely on the defensive and reliant on the grain-producing areas they controlled, while still able to mount attacks on Japanese communications. The net consequence was a stalemate until fresh Japanese offensives in 1942 and, still more, 1944–5 altered the situation as well as displayed Japanese capabilities; in the meanwhile, this stalemate put a major burden on the Japanese.[14] Separately, the Soviets proved better able to meet the necessary logistical challenge of deploying large numbers and fighting in the Far East, as shown in their battle with the Japanese at Khalkhin Gol in 1939,[15] and, at a far greater scale, in their conquest of Manchuria in 1945.[16]

Very difficult mountainous terrain was a more serious issue for the Italians when they conquered Ethiopia in 1935–6, and the logistical problems of campaigning forced the Italians, despite their experiment with air-dropping live animals to supply their troops with meat, to devote much energy to road building and led to commentary on the campaigning to focus on the availability of good roads that could be used for mechanised columns. The Italians certainly suffered more than they had anticipated in their planning from poor communications and the limited local availability of supplies. The British War Office reported in January 1936:

> Owing to the check on the Tembien plateau and the delay in starting construction of weather-proof lines of communication, it now seems improbable that the Italians on the northern front will be able to advance much further, or secure a decisive success, before the opening of the wet season in April puts a stop to active operations for a period of about five months…. short of killing the Emperor of Ethiopia or engaging in large scale gas warfare from the air, it is difficult to foresee what other major success the Italians can now hope to achieve.[17]

In the event, a larger and regrouped Italian army, backed by the use of gas, as well as the crucial response of the opponent, in this case the Ethiopians choosing to engage in battle rather than rely on guerrilla warfare, ensured that Italian success proved easier than British and French military commanders had anticipated. The Italians benefited from the British not intervening by applying, as considered, oil sanctions or closing the Suez Canal to Italian resupply ships, a step that would have greatly harmed Italian logistics. Italy lacked oil, whereas Britain controlled the production of oil in Iraq and Persia (Iran), as well as its shipping.

Alongside Italy's deficiencies, the pace of advance was far quicker than it had been for them in the 1890s, the high tempo owing much to logistical improvement thanks to extensive road building and to the resulting Italian mobility enabling them to retain the initiative. Once established, a resistance of ambushes and surprise attacks on precarious supply routes was countered by building forts, the recruitment of local troops, and savage repression. As far as the latter was concerned, the French crushed nationalist uprisings in Vietnam in 1930–1 in part by the counter-logistics of killing livestock, village burnings and crop destruction.[18] In counter-insurgency operations, terror was often applied, and by both sides, as a consequence in part of the difficulty of enforcing control in a situation in which poor logistics led to a precarious presence.

A very different war, the conflict between Bolivia and Paraguay in the Chaco War of 1932–5, underlined the logistical implications of the physical environment. A British officer reporting from the Bolivian side noted:

> Paraguay, deriving advantage from her geographical position, is placed comparatively near the scene of operations, and her troops, supplies, and food are transported to the front by a combination of light railway, lorry, mule, and march methods. Bolivia, on the other hand, suffers a separation of 1,000 miles between her forces and her main base at the capital, of which only 500 miles are covered by rail. Over the remaining distance runs a narrow and broken road, at first over high rocky plateaux, and then over wooded mountain slopes down to the flat plains of the Chaco jungle. This is thick with dust in the winter and often washed away in summer. Troops and supplies perform the 500-mile trek by lorry and donkey, while in the opening

stages of the war, when transport was extremely short, some of the regiments were compelled to march most of the way on foot, ankle deep in the dust and suffering torture from thirst and danger.[19]

In the harsh, largely waterless, scrub terrain beyond the humid flood plains, the Bolivians lacked the water necessary for their larger army and its tactics of the mass offensive. The Paraguayans were reduced to mules and oxen for transport and lacked sufficient arms, ammunition and medical supplies, but understood how to fight in the terrain, an ability important to manoeuvrability and logistics.[20] The situation was even more difficult in the distant tropical rainforest in which Columbia and Peru clashed in 1932–3. With Brazilian permission, Colombia sent a naval flotilla via the Atlantic and Amazon, a reminder of the extent to which naval power could also involve inland waters, and a matter of both force projection and logistical support in one.

In the Spanish Civil War (1936–9) both sides were inadequately supplied, while the poor training fed into a limited understanding of logistics, and notably so on the Republican side, where the worker militia were especially short of equipment, although, as a result of pre-war planning, the defence committees were able to provide organisational strength, including in logistics, in 1936–7.[21] Both sides improvised, Franco seizing buses and trucks alongside mules and oxen to support his drive on Madrid in 1936. J.F.C. Fuller, a retired British general who was with the Nationalists as a journalist, sent a report to British Military Intelligence in March 1937 including:

> … had Franco a highly organised army and plenty of transport he could take Madrid. But he has not. For instance, General Queipo de Llano told me himself that, when he launched his advance against Malaga, he had only 28 lorries…[22]

That month, an Italian advance east of Madrid towards Guadalajara became overly dependent on the few roads in the region, lost momentum and logistical support in poor weather, and was finally driven back in a successful counter-attack. In April 1938, the British Assistant Military Attaché in Paris commented, after visiting Nationalist Spain: '… comparatively small forces are strung out on a vast length of front…. an incomplete and ad hoc organisation.' Poor transport and roads were discerned, as well as the conduct of the war 'in an utterly haphazard way'.[23] This was overly harsh, as there was an outcome, one to which the Nationalists better logistics contributed. Nevertheless, logistical deficiencies were aspects of more general issues with organisation and, in particular, at the operational level of war.[24]

A very different environment was to prove logistically highly challenging for the Soviets when they attacked Finland in 1939–40. There was a shortage of winter wear, which was closely linked to numerous casualties from frostbite, and also affected morale. The Soviets were also affected by a shortage of lorries, in part due to having few at the beginning of the war, but combat, wear and tear, and the impact of the cold were also important. The shortage of lorries restricted deliveries of food, fuel and ammunition, while the alternative, horse-drawn transport, was made less viable by the cold killing or disabling horses. For humans, let alone horses, rations were limited, no water was provided, and the use of melted and boiled snow was linked to illness. Cooking was difficult, and the Soviets relied on hardtack biscuits. Yet, despite this, the Soviet army persisted, fought hard, and won,[25] which looked toward its more hard-won achievement in The Second World War.

Chapter 9

The Second World War, 1939–1945

The late 1930s was one in which just-enough logistics was the norm, as with German rearmament, which, indeed, had a 'shop window' character, as was seen with the mobilisation at the time of the Munich Crisis in 1938. This mobilisation revealed shortages in military stores and indeed all the way back along the supply chain as lack of sufficient raw materials, notably iron and steel, hit armaments' production. In turn, this production was to be greatly helped by the German seizure of Czech armaments, industry and raw materials in March 1939 doubling the quantity of artillery the Germans controlled. In addition, the strength of German armour when Poland was attacked in September 1939 owed much to Czech tanks (appearing as P3Kw35s and 38s in Poland and France in 1939–40) and military-industry capacity. The seizure of assets has always been significant in logistics, and was expanded in the war, notably with that of airfields. The use of Czech territory altered another aspect of the logistical paradigm, by making it possible to attack Poland from even more of its perimeter. However, in terms of resources, there was not the oil to meet Göring's bold plan to take the *Luftwaffe*'s strength to 21,000 aircraft by the end of 1942. The German shortage of oil also made the plans for a much larger tank force implausible, while there was neither the shipbuilding nor dockyard capacity for the naval construction programme that was endorsed in the winter of 1938–9, nor indeed the steel or strategic minerals such as tungsten.

Blitzkrieg was never an official German term but rather a description of adding an armoured and airpower leading component to very traditional German approaches to war. In practice, the *ad hoc* nature of *blitzkrieg* helps put supposed German operational brilliance in 1939–41 in its proper context, as does the extent to which it arose from sequential war-making and one-front campaigning. Moreover, much of the German army in 1939 was heavily reliant on railways and draught animals. The potential of weaponry and logistics based on the internal combustion engine, notably in powering tanks, was less fully grasped than talk of *blitzkrieg* might suggest, not least because, despite the popular view that the German army was largely mechanised, much of the German army was unmechanised, heavily dependent on horse-drawn supplies, and walked into battle. *Panzergrenadiers* (mechanised infantry) were only a minority of the German infantry. The absence of adequate mechanisation, at nearly all levels, reduced the effectiveness and range of German advances; although, even had there been more vehicles, there were the issues of their maintenance and, more seriously, or the availability of oil. In light of the latter factor, the question of 'fitness for purpose' occurs anew.

Problems were to be revealed in the German invasions of Poland (1939) and France (1940). In the first, there was reasonable supply provision in the preparation stage, with food, ammunition and fuel, but fuel shortages were encountered by the third day of the invasion and there had been no practice in the large-scale resupply of fuel, nor indeed much in vehicle recovery nor field repairs. Thus, it proved difficult to sustain the operational tempo of mechanised warfare. This repeatedly proved a problem for the Germans, not least due to scant mobile logistical capability, an issue that was particularly problematic with fuel and maintenance.[1] Support doctrine and practice were inadequate. These issues were to recur for the Germans in 1940. Linked to these, the German *panzer* advances required periodic rests.

Logistical failure was an aspect of the serious flaws in German war-making in the invasion of the Soviet Union launched on 22 June 1941, flaws, paralleling those made by the Japanese in China in 1937, that reflected major limitations in German operational thinking and strategic planning.[2] In strategic terms, the invasion of a country in order to obtain the resources that would have enabled your invasion to succeed was problematic, the seizure of food certainly alienated the civilian population, and there was no goal short of absolute victory, which left no basis for a compromise peace. In military terms, the scale of the Soviet Union was a major problem because of the range and speed of the necessary advance once the Soviets did not collapse, as anticipated, in initial frontier battles. This scale added to the strain on existing stores without providing a substitute of gaining the Soviet ones that might have been obtained had the invasion been mounted after the harvest. Indeed, an attack in mid-summer posed particular problems for food stores, although grass was available for draught animals. However, an attack later in the year would have lessened the campaigning opportunity before the onset of winter.

The range, speed and poor roads also put pressure on German equipment, which focused attention on the deficiencies of spare parts logistics. The variety of vehicles in the German army exacerbated this issue which itself became more difficult due to distance from parts-depots. The improvised character of German logistics, indeed war-making, not least in the absence of accurate Intelligence about the Soviet Union, was obviously inadequate even before the onset of winter. Damaged by the fighting, with, in particular, the bridges over the River Dnieper destroyed by the retreating Soviets, the Soviet railway system when partly taken over by the Germans posed major difficulties in terms of the necessary re-gauging with the change from one track gauge to another. In addition, lorries were too few in number and many requisitioned ones broke down, including large numbers of those seized in France. As a result of these transportation problems, there was grave difficulty in meeting supply requirements, for example, of winter uniforms.

These and other problems interacted in a context in part defined by a general failure of Intelligence about the situation in the Soviet Union, from the difficult terrain and lack of forage and roads on the Arctic front toward Murmansk which the Germans failed to capture,[3] to the myriad problems for vehicles caused by summer dust, autumn mud in the *rasputitsa* (time without roads), and winter snow. These logistical problems were not so much a product of deficiencies in the abstract but, rather, that of a total failure of fitness for purpose, one that was a reflection of strategic and operational inadequacy on the part of the German command. In North Africa, moreover, Erwin Rommel, the Axis commander, paid insufficient heed to logistics. Similar issues, especially the mismatch between assumptions and capabilities, were to affect Soviet counter-offensives, particularly in early 1942 and early 1943.[4]

The inadequacies of German logistics on the Eastern Front, and the related need for food, one that came into greater prominence from August 1941, played a minor role (in comparison to racism) in contributing to the genocidal determination to kill Jews and the murderous treatment of other Soviet citizens.[5] In turn, logistics played a role in the treatment of Jews who were not at once slaughtered. This was because of the use of slave-labour as at Auschwitz, where the nearby large I.G. Farben plant was the largest German producer of synthetic rubber and oil, both crucial for the war economy.[6] So too, more generally, with millions of prisoners of war and foreign workers forced to work for the Germans, and receiving low-calorie rations.[7]

The Germans were not alone in facing logistical difficulties on campaign. The Italians in Greece in the winter of 1940–1 found their supplies disrupted while the logistical problems of their Greek opponents included a shortage of shells. These problems were to affect them anew when attacked by Germany in April 1941. In contrast, experience in the logistics of distant operations helped the British. Thus, the British conquered Italian East Africa (Ethiopia, Eritrea, Italian Somaliland) in 1941 after overcoming formidable logistical difficulties and not only those posed by the environment.[8] The problems they faced in the region, however, were to a degree eased by their logistical back-story, in that the wider oceanic and imperial spaces abutting and leading to Italian East Africa were under British control, which permitted the management of logistical preparation. In particular, the nearest hostile forces, those of Italy and Germany in the Mediterranean and North Africa, and of Vichy France in Madagascar and the Near East, were not in a position to intervene. The British benefitted from the roads built by the Italians in recent years.

Despite major logistical limitations of which they were aware,[9] the Japanese rapidly conquered a major area, from the Aleutian Islands to Burma and the South-West Pacific, in late 1941 and early 1942. They benefited from the ability of their navy to operate large-scale, underway refuelling, a process that contributed greatly to their capacity to take the initiative, notably with the attack on Pearl Harbor in 1941, and the mission to the Indian Ocean in 1942.[10] At that stage, the Americans could not match this capability, and that did not change until a massive expansion in American shipbuilding and the development of refuelling skills.

The sequential nature or leapfrogging of Japanese amphibious operations reflected the need to move forward with bases, for both ships and aircraft. The physical infrastructure in the Dutch East Indies (Indonesia) was largely a matter of ports, which was a consequence of its role as a sprawling archipelago. Logistical power-projection capability was therefore directly linked to the ability to seize and guard strategic shipping straits, such as the Sunda Straits between Java and Sumatra. Air support was the crucial adjunct which helped close these waters to Allied shipping.

There were major differences in tasking and therefore logistical requirements between the countries in each alliance system. Nevertheless, in aggregate terms, in contrast to the Axis powers, the Allies faced a greater range of tasks, notably including strategic bombing and large-scale amphibious operations, but, more generally, were in a better situation. American military capacity became particularly important, and in all aspects of operations. Thus, American infantry and artillery were fully motorised, which helped maintain the pace and cohesion of the advance. The Germans, Japanese, and Soviets, and even the British, could not match this integration; although that did not prevent major Japanese advances in China in 1944–5. Moreover, although the British were semi-dependent on American vehicles, the British Eighth Army and Rommel's *Afrika Korps* in 1942 waged desert battles without reliance on horses. Later, the whole of the British 21st Army Group in 1944–5, from D-Day to VE-Day, was fully motorised.

American force structure and tactics were a direct product of the economy's ability to produce weaponry and vehicles in large and unprecedented numbers, and were closely related to American logistical capabilities. The force structure and logistics also helped ensure the strength of the economy, as the relatively small number of combat divisions, eighty-nine, made it easier to meet demands in America for skilled labour. President Franklin Delano Roosevelt's call, in his radio broadcast on 29 December 1940, for America to be 'the great arsenal of democracy' was fully met. Indeed, the Americans also provided crucial support to the British war economy, from munitions, such as 5,760 tanks, to raw materials, such as synthetic rubber, oil and steel.[11]

Given the size of the United States, it was necessary for the economy there to be effective in transportation and logistics, an experience that helped greatly in the war, while the flexibility of both economy and society had direct consequences in terms of production and fighting quality.[12] Institutional and cultural factors were very significant, notably the widespread existence of problem-solving empirical practices and management abilities stemming from the needs of the economy, as well as a high degree of appointment and promotion on merit, albeit not as far as African-Americans were concerned. In addition, there were widely-disseminated social characteristics that had military value, including a can-do spirit, an acceptance of change, a willingness to respond to the opportunities provided by new equipment, a relative ease with mobility, and a self-reliance that stemmed from an undeferential and meritocratic society.[13] At the same time, the Americans needed first to build a global strategic transport system, which was why the idea of invading France in 1942 and 1943 was not practical.

At both tactical and operational levels, mobility and firepower were seen by the Americans (and, separately, British) as multipliers that compensated for deploying relatively few troops, in part due to the requirements of the home economy and in part due to the manpower, not least high-grade manpower, used for air and sea warfare. These facets had resource and logistical implications, not least in the need for oil (gas), ammunition, and shipping. Thus, the *relatively* small size of the American combat arm increased its mobility, although there was need, on the part of Americans, for a substantial backup support system (and doctrine accordingly) and for a higher level of resources than enjoyed by other armies. The greater the availability of mechanised transport, the more it was employed.

In contrast, Germany, the Soviet Union, Italy and Japan all made great use of horses. The Germans had 625,000 horses ready for the assault on the Soviet Union in June 1941, but the death rates were high, as they were even more for the Soviets. Mules were also important for the British, and notably so in North Africa, Burma and Italy. The Americans also used mules, including in Italy, the South-West Pacific and Burma, but, if the slopes proved too difficult for them and they slipped too frequently, then human porterage was necessary.[14] In Burma, both Britain and Japan used elephants. As with machines, the employment of animals in turn ensured logistical requirements, not only of food and water (fuel), but also maintenance, particularly with veterinary support. In order to provide mobility and save on fuel, the Germans made extensive use of bicycles, manufacturing types able to carry fully-equipped soldiers; but these bicycles were dependent on roads and provided no protection against enemy fire.[15]

The emphasis on machinery ensured a reliance on oil, and its supply became a key element of wartime logistics. Conversely, the lack of oil was a major German weakness that was accentuated by Allied air attack.[16]The Anglo-Soviet occupation of Persia (Iran) in 1941 was in part in order to provide an Allied route to the Soviet Union, but was also seen as the basis for British air attacks on the important Baku oilfields in Azerbaijan, then part of the Soviet Union, if the Germans seized them.[17] Similarly, the capture of southern Italy in 1943 provided a base for Allied strategic bombing including, in 1944, heavy raids on the oil refineries in Ploesti, Romania. The Italians had been less successful in 1940 in their attack on British refineries in Haifa and the Persian Gulf.[18] A very different counter-logistics of oil was seen with the German summer offensive of 1942 which was not only designed to secure the oilfields of the Caucasus for the Germans, but also to cut the supply of oil to the Soviets.[19]

The lack of oil meant major problems, as with German tank operations in North Africa in 1942, or the shortage of *Luftwaffe* training in the latter stage of the war and the resulting decline in fighting quality. Japan was also hit hard by a shortage of oil. So also with the reliance on rubber for tyres: logistics was both global in scale and comprehensive in character. Alongside the significance of oil, however, came the continued role of coal,[20] not least for railways but in practice for most of industry.

Although effects-based air operations proved difficult to evaluate because of issues in understanding the (anyway changing) German war economy,[21] air attack on German communications and thus logistics was crucial to their capability, largely because it was eventually done on such a massive scale and because the targets could not be attacked by any other means. The reliance of logistics on rail was far greater than today, and that reliance increased its vulnerability to attack, because rail systems lack the flexibility of their road counterparts, being less dense and therefore less able to switch routes. As critical points, bridges, for example over the Seine in 1944, marshalling yards, for example the German one at Hamm near the Ruhr in 1943, and railway engines, proved particular targets for attack. Allied bombing brought the SNCF, the French rail system, then under German control, to collapse, followed by the *Reichsbahn*, the German rail system.[22] Damage was extensive enough to preclude effective repairs, which indicated the potential for increasing returns to scale in the air offensive. Its vulnerabilities ensured that the rail system was also prone to partisan (guerrilla) attack in occupied areas. In turn, the Germans devoted massive efforts to the protection of their transport system, notably with many guns in antiaircraft positions.

The importance of logistics meant that attacks on supply lines were highly significant, and at every level from the tactical up. Thus, in

Burma, the Allies operated in part by threatening Japanese supply lines. Roadblocks were part of the system, a means of forcing the Japanese to retreat or attack. In turn, the Allied forces were in part dependant on air supply which was enhanced by the provision of forward airstrips that were created in part by bulldozers brought forward by glider.[23] However, the weather affected flying, and thus, on a continuing pattern, the monsoon remained a major challenge.

The Allied superiority in logistics was repeatedly demonstrated, and against Germany, Italy and Japan. In the case of the last, Chinese logistics were poor, but American (in the Pacific), Australian (in New Guinea), British (in Burma), and Soviet (in Manchuria) all proved superior; the Soviets benefiting from their attack being mounted at the very close of the war.

In the Pacific, the American problem-solving, can-do, approach to logistics permitted a rapid advance. The 'fleet train' that provided logistical support was greatly expanded, while processes for transferring fuel, ammunition and other supplies from ship to ship while underway were developed, although it was post-war that 'one stop vessels' able to transfer all supplies were developed.[24] In addition, the use, from 1944, of shipping as floating depots for artillery, ammunition, and other *matériel* increased the speed of army resupply, as it was no longer necessary to use distant Australia as a staging area for American operations. The American advance across the Pacific would have been impossible without the ability to ship large quantities of supplies and to develop the associated infrastructure, such as harbours and oil-storage facilities, and build the large number of ships of the support train. In some respects, this was a war of engineers, notably in the Pacific, and the American aptitude for creating effective infrastructure was applied to great effect there, with the ability to mount both a Southern drive and a Central Pacific one. Cumulative experience helped in naval logistical support, both at sea and with harbour-bases, but also was the legacy of the problem-solving interwar leadership development based at the war colleges.

Geopolitical circumstances were important as seen in particular with the contrast between the American ability to use the eastern Pacific as a large, but safe, transit zone for providing supplies further west, as opposed to the problems faced by the Germans and increasingly by the Japanese. German ships, both surface and submarine, were especially vulnerable in getting to the space for operations, not least because they could be engaged on the way to or from their logistical bases. Thus, submarines were attacked by Allied aircraft while they crossed the Bay of Biscay from their French bases. The availability of more aircraft by late 1943 meant that submarines could not sail safely on the surface and thus used up some of their supply of electricity, as the recharging of batteries had to be done on the surface which made submarines vulnerable to air attack. In sailing on the surface, the submarines operated using diesel oil in engines that could drive the vessel three to four times faster than underwater cruising. This factor affected cruising range and the number of days that could be spent at sea, yet again demonstrating the extent to which the effectiveness of weaponry in part depended on the anti/counter-weapons and tactics. Operating in the mid-Atlantic, the Germans had Type XIV supply submarines, nicknamed 'milk cows', able to supply fuel, torpedoes and fresh food, which enabled U-boats to remain off the American coast and operate in the Caribbean. Ten Type XIVs entered service but, a focus for Allied attack, to which they were vulnerable, all were sunk, the last in June 1944.

During the war, scale as a key element of capability ensured a range of logistical requirements. General Ismay, in effect his chief of staff, informed Churchill in March 1944:

The war we have to wage against Japan is of an entirely new type. It is no mere clash of opposing fleets. Allied naval forces must be so strong in themselves, and so fully equipped to carry with them land and air forces, that they can overcome not only Japanese naval forces but also Japanese garrisons supported by shore-based air forces.[25]

This was full-spectrum logistics, and it required unprecedented shipping support and, therefore, shipbuilding. Both were magnified by the distance of America from combat zones and from the range of zones in which the Americans engaged.[26] Mass production of standardised designs, with the work carried out by well-remunerated workers free from the threat of bombing, were all crucial factors that it is easy to run together without appreciating their individual and collective significance. So also with the dependence of logistics on manpower, a factor seen across time. Thus, this merchant marine required a large number of seamen, many of them highly trained.

Yet, at the same time, across the spectrum, and for both sides, appropriate and rapid resupply was a major issue, one affected by the range of operations and by transport problems, as for both sides in the Soviet Union. At the same time, logistical effectiveness rested on a range of capabilities, including Intelligence superiority, with broken codes helping the Allies attack the Axis resupply of their forces in North Africa, which contributed to the poor logistics that hampered the Axis there. At El Alamein in 1942, the well-supplied British artillery owed much not only to preparedness but also to Intelligence superiority.

There was improvisation by the Allies, not least in the face of difficult terrain, as in Burma and New Guinea, but also marked improvement. In part, this was a matter of the application of resources, as with the use of air-dropped supplies, notably by the British in Burma. The integration of supplies into doctrine at all levels was also significant, and contrasted with the weak state of Japanese logistics in their attack on the India-Burma frontier in 1944. The Japanese troops carried seven days' rations, took with cattle, and relied on mules and oxcarts for transporting supplies, but, more particularly, on seizing the British supply base at Imphal. Held off by British forces at Kohima and Imphal, forces who were resupplied by air, the Japanese ate their mules and oxen, and retreated with many casualties. The Chinese troops in Burma also had simple logistics, with porters playing a major role.

The Japanese logistical practice of living off the land in heavily-populated China was far more difficult when extended to the Central Pacific and South-West Pacific theatres and highland Burma, where even the basic sustenance, to say nothing of hygiene and medicine, were lacking. Yet, units were sent to the front in the Solomons, New Guinea, and Indonesia, with almost no supplies, over and over again, trying to live off the native population who were in a subsistence economy themselves. There was also a logistical catastrophe for the Japanese troops sent to the Aleutian Islands in June 1942.

German doctrine and warmaking also proved seriously flawed in the case of logistics, and did so compared to all Germany's opponents. At the systemic level, relationships within the Nazi system bound parts of the economy to sections of the military, so that, for example, the army and the SS competed for resources in part in terms of these links. This situation prefigured the later links of militaries to the economy in many states.

Allied logistical strength was especially apparent in the case of the Americans and, albeit to a lesser extent, British, and rose to the fore from the well-prepared and fully-resourced D-Day operation on 6 June 1944 onward. That was a combined and joint operation in every sense, and worked as such, not least due to effective commanders, especially John Lee, the logistics supremo.[27] Yet, the Allies suffered greatly from the determined resistance the Germans mounted in the ports and with the destruction of harbour facilities, especially by the Germans at Cherbourg and in Allied bombing at Le Havre, where most of the docks were destroyed. As a result, unloading capacity in France remained a major issue, although Cherbourg's port was brought into partial use by mid-August, while Le Havre was open again by 9 October. However, although Calais fell to attack on 30 September, Dunkirk remained under German control until the end of the war, in part because of a decision to focus efforts on Antwerp. The facilities at heavily-damaged Calais were not usable until November, and only then for personnel. Allied air attacks had also devastated the rail system, particularly the bridges. The latter required much repair and that took much time.[28]

In what Rommel had warned would be a *Materielschlacht* (battle of supply), the campaign in France saw Allied logistical mastery in Normandy, in part by supply across the beaches through the 'Mulberry' prefabricated harbour, but also a difficulty in applying and sustaining resources, with port capacity remaining a key issue. The capture of Marseille and Toulon immediately after Operation Dragoon, the Allied landings in Provence in August 1944, made a huge difference to Allied logistics that autumn when the Scheldt estuary had yet to be cleared of German forces. This capture meant that the US 7th and French 1st Armies (the 6th Army Group) had its own LOC separate and independent from those of 21st and 12th Army Groups.

The Allied broad-front approach permitted a drawing on a variety of routes, but also increased the logistical burden of supporting the advance. The Red Ball Express lorries were effective in speedily moving the supplies made available, but the distance to be covered was a serious issue. Moreover, the failure to clear the Scheldt forced advancing Allied units to remain dependent on overlong supply routes as they neared Germany, which caused major problems despite their degree of mechanisation; whereas the Germans, as they retreated and consolidated, benefited from shorter routes.[29]

The Soviets also faced operational and tactical logistical strain, as in the unsuccessful attack on Belarus in October 1943 and the invasion of Romania in early 1944. However, aside from plentiful resources from the factories, notably new tanks, the Soviets proved good at using the resources they had, ranging from tanks that could be repaired when they broke down, to troops and horses that could be expected to fight with little food. At the strategic level, Soviet logistical capability was badly affected by the seizure of so much productive agricultural land by the Germans in 1941, notably Ukraine, but, conversely, benefited greatly from the movement east of so much manufacturing plant that year. The consequent need to create new supply systems, both for and from the new factories, was one of the major logistical successes of the war, one not matched by the Germans, although the latter benefited from the food and prisoner-of-war labour seized from their conquests.

By 1945, Allied logistics was very slick. Eisenhower had always foreseen a major battle to cross the Rhine and enter Germany. He knew that he would need a huge logistical machine because the Germans would provide nothing, instead relying on destruction and scorched earth counter-logistics, and he was correct to do so. Whereas the battle for Normandy was far harder than had been anticipated by the Allies (with logistics proving a major problem for all armies), and both the Arnhem offensive and the response to the German Ardennes attack, the Battle of the Bulge of December 1944, were unplanned, the advance into Germany was anticipated and planned for, which helped ensure that the 1945 campaign went so well. In addition, the experience of the Ardennes offensive meant even greater Allied flexibility than hitherto. Moreover, the southern French ports (supplying up to 25 per cent of Allied needs) continued greatly to ease the logistics flow. The speed of the advance in 1945 was eased by the Germans being in no position to contest control of their own airspace. In addition, the plans of Ira Eaker, the commander of the Eighth Air Force (8AF), for a focus on oil and transport had bitten deep into the German ability to manoeuvre. Allied military manufacturing was running at full capacity, and to the primary benefit of the Germany-First theatre. Sequential logistics were planned: the assumption was that the European war would soon end, and then forces and logistics would switch to the Pacific, with a large-scale invasion of Japan in 1946.

Highly important as it was, the key to winning the logistics war was not industrial capacity per se, but the supply, training and maintenance systems that established effectiveness,[30] and where you pointed the finished product. The emergency of the German Bulge offensive brought out the best of American improvisation. They used all their 6x6 lorries and Antwerp-based articulated lorries (semi-trailers) to shift first men (standing up, in huge numbers), then their kit and supplies, to where they were needed.

For the subsequent push into Germany in 1945, this flexibility remained, due to the luxury of sheer quantity and the commonality of parts of a relatively small range of weapons. The American 101st Airborne Division was able to take DUKWs with them into southern Bavaria, as far away as possible from the ocean shores in which they were designed to operate, because they could. Resupply, a key element, was more successful than during the Allied advance in 1944, and the advance was eased by a determined engagement with the degree to which the Germans would seek to wreck the transport infrastructure. Engineering units, organised as General Service Regiments, played a major role in addressing the situation, not least in bridge-building.[31] The biggest brake on the Allies in Germany in 1945 was not logistics, but manpower. Their other logistical failure was the inability to cope with over one million German prisoners.

Alliance cooperation was also important, both in resource development and production,[32] and in supply systems. By 1 July 1942, 16 per cent of Soviet tanks were foreign supplied, mostly British.[33] Weapons were not the sole issue. By 1944, the Soviets were able to make much of their infantry and artillery mobile, in part thanks to the American provision of lorries.

A good example of logistical cooperation was, as ever, provided by basing units abroad. For the American air force in Britain, the 8AF, a significant logistical support infrastructure came from the RAF and British ministries. Airfields, depots (supply and maintenance), port facilities, warehouse, and POL facilities, civilian labour (at airbases), communications, training areas, security (air defence), and other features provided the geographic logistic infrastructure into which the American supply system could flow. So it was not what is called today an austere environment into which to deploy an expeditionary air force. Moreover, many operational and equipment innovations the 8AF experimented with were provided by British industries, including some of the airborne radars and the Disney rocket bomb. Thus, technology transfers and information sharing, such as aerial reconnaissance photography of targets, was important, just like managing ground radar and radio frequencies between the RAF and 8AF which was a form of logistical support. Road coordination to move ammunition from the port of Liverpool to 8AF bases by the 1511th Quartermaster Trucking Regiment that directly supported 8AF was an important part of the logistical system which focused on the 8AF logistics command that ordered, received and distributed supplies. The main supply depot was at Warton near Liverpool, with the HQ at nearby Burtonwood, but there was a series of depots and truck transport stations across Britain. As the 8AF increased its number of units and the size of its footprint in Britain, there was a corresponding reorganisation of the supply side. Adding complexity, it was also necessary for the US forces to consider the theatre level logistical echelon.

Road movement was important to the Allies, but, if the German reliance on railways was not matched for the Allies, it was still highly significant, and not only for the Soviets. The United States Military Railway Service ran operating and shop battalions,[34] notably taking responsibility for the maintenance of rolling stock near the front line, as with the 756th Railway Shop Battalion when France was invaded in 1944.

As in other forms of transport, specifications were different to those of the late nineteenth century. Thus, railways were required to carry larger, heavier and different loads, while the expectation of ship unloading facilities had altered, leading to the Allied reliance on the Mulberry Harbours and buoyed pipelines from tankers in the Normandy campaign, as well as the PLUTO undersea pipelines moving fuel from Britain to France. The problems facing Italy in supplying Libya across the Mediterranean also to a degree were related to harbour facilities.[35] In contrast, Suez was able to handle far more British supplies while in 1944 the Americans were able to deliver large quantities to North Africa as well as the locomotives and wagons necessary to increase dramatically the capacity of the rail system in Algeria.

At the same time, the value of particular systems in part depended on the nature of opposing action. Thus, in resupplying besieged Tobruk in 1941, the British logistics were in part defined by the Axis counter-logistics: the British found large freighters particularly vulnerable to Axis air attack, instead coming to rely on faster destroyers. British ships were also particularly exposed when unloading to lighters, but, on the other hand, the Axis lacked the necessary aircraft armed with cannon and lacked sufficient air-dropped mines. Thus, logistical possibilities had at the time to be considered across a range of possibilities, as remains the case.[36] Holding Tobruk also denied it to the Axis forces, thus greatly extending their supply chain.

One factor that tends to be underplayed during the Second World War, due to the role of conscription was that of payment for the military. That branch of logistics, however, continued to be significant, although in a variety of ways, some of them unexpected. Hitler providing funds, including estates, to military commanders was one such. So very differently, with the provision of payment, for soldiers and their families, to Britain's Indian Army, a significant force of volunteers.[37] Logistical support for that army improved with time which in part was a reflection of the scale of the effort required, but also of the need to reconfigure a military primarily designed to fight on the North-West Frontier for the very different terrain, vegetation and challenge of war with Japan on the Burma frontier. As part of that reconfiguration, food and medical supplies improved, and there was also a response to the requirements of the troops, notably providing, in place of dehydrated meat and tinned food, recently-killed animals that had been transported live from central India from late 1943.[38] If that was one reminder of the range of logistical requirements, the significance of postal deliveries to military morale created another.

Again, however, and unsurprisingly so, the situation varied greatly by country, not least in terms of how morale and its maintenance were understood. So also with the tasks, with America and Britain fighting industrialised war in jungles and on remote islands, a formidable task requiring not only industrial mobilisation but the ability to develop and sustain long-distance and large-scale supply chains.

Chapter 10

The Cold War and Beyond, 1945–2021

T he logistical environment did not change appreciably in the second half of the century. There was the overhang of atomic power, but the key range of conflict continued to be pertinent and the major contrast was the absence of direct warfare between the leading powers. Both the Americans and the Soviets had to have the logistics in place to wage World War Three, and Soviet planning was based on having the supplies for a rapid advance across West Germany and into France, but that war did not occur. Instead, each power found itself involved in counter-insurgency warfare, notably the Soviet Union in Hungary in 1956, Czechoslovakia in 1968, and, very differently, Afghanistan in 1979–89, and the Americans in the far-more-distant Vietnam War.

As far as land conflict was concerned, there was a great emphasis on mobility which was regarded as necessary because it was assumed, by both sides, that any war touched off by a Warsaw Pact invasion of West Europe would be rapid. Putting an emphasis on mobility and tempo, the Soviets planned a rapid advance into NATO rear areas, which would compromise the use of Western nuclear weaponry. Essentially building on the operational policy of their advance in the latter stage of the Second World War, with its successful penetration between German defensive hedgehogs, the Soviet Army put a premium on a rapid advance.

However, the Soviets would also have faced logistical stress. With the revocation of wartime American Lend Lease aid, the Soviets were no longer receiving spare parts for their military motor transport, which was composed overwhelmingly of Ford and Studebaker lorries supplied by the Americans during the war. The equipment for routine tune-ups, such as spark plugs and distributors, and for oil changes, as well as other *matérial*, such as batteries, tyres, inner tubes for tyres, oil and air filters, were no longer available. Nor were axles, drive shafts, gearboxes and engine blocks. Tanks alone rolling ahead would not have been enough, and the horses sequestered during the war for the army had to be returned as Soviet agriculture was in a bad shape. Yet, the Soviets could still have advanced far before they hit logistical crisis. Indeed, predictions that they could only advance for 6–9 days sometimes ignored what they could have achieved in that period.

In turn, a flexible defence was called for by Western strategists, not least with counter-attacks to take advantage of the use of nuclear weaponry. Furthermore, only a flexible defence would allow the Western forces to regain and exploit the initiative. The skill of the Israelis against Arab defensive positions in the Six-Day War of 1967 and, eventually, after initial Egyptian and Syrian successes, in the Yom Kippur War of 1973, appeared to show the vulnerability of forces with a low rate of activity. The emphasis on manoeuvre and high tempo operations by both the Soviet Union and NATO helped lead to a greater mechanisation of logistics.

Lorries became even more important to logistics because of the situation outside Europe with the end of the large-scale use of the horse, the spread of oil supply facilities, and the extent of campaigning where rail networks were sparse. Moreover, the very emphasis on insurrectionary and counter-insurrectionary warfare meant that operations were generally conducted in difficult terrain and economically marginal areas, which again put an emphasis on road, not rail, transport. However, this emphasis was lessened because insurrectionary movements frequently used human porters, notably along the Ho Chi Minh Trail in Southeast Asia during the Vietnam War; while counterinsurrectionary forces enhanced mobility and lessened vulnerability to ambushes by employing air-drops and helicopters, again particularly during the Vietnam War, but already earlier, as by the French in Algeria.[1] Aerial supply had also been used by the Americans in Korea.

The significance of porters to insurgent forces ensured that the attempt to prevent this supply was part of the counter-insurgent strategy. In the case of Malaya in the 1950s, the British were able to cut the insurgents off from their Min Yuen porter support due to the success of compulsory resettlement in New Villages. The Americans, however, in the Vietnam War, fighting a larger-scale insurgency, could not quite replicate this in the Strategic Hamlets programme and in the attempts to interdict the Ho Chi Minh Trail.

Aerial resupply exemplified the extent to which logistical methods, platforms and doctrine depended in part on the relative advantage over anti-methods. Thus, in the Vietnam War, land warfare became more mobile as a result of the helicopter but in part only because the North Vietnamese did not have human-portable surface-to-air missiles until late in the war. Had they done so earlier, the usage of helicopters would have been extremely difficult which would have presumably forced the Americans to change their tactics to more conventional methods of advance, supply and retreat. This problem affected the Soviet forces in Afghanistan in 1979–89 in a conflict in which the harsh terrain exacerbated major logistical problems. In both conflicts, the need to control territory, and not only cities, in order to influence the population[2] posed particular logistical challenges. So also did external intervention from safe bases: the United States, the Soviet Union and China in the Vietnam War, and the first two in the Afghanistan conflict, with the Americans supplying anti-aircraft missiles to the Afghan opponents of the Soviets.

Airlift was more generally a matter of operational and even strategic value, as well as tactical, during this period. Thus, tactical usage was shown in the British reliance on helicopters to supply fortified posts near the Irish frontier during the Northern Ireland 'Troubles' from the 1970s to 1990s, thereby avoiding ambushes of road convoys. This was significant, but so also were strategic factors, notably British control of the sea and the use of Intelligence both serving to prevent heavy arms reaching the IRA.

Strategic capability was displayed in the 278,228 Anglo-American flights that provided 2.3 million tons of supplies to enable Berlin to thwart Soviet blockade in 1948–9. Airlift also overcame maritime blockade, as with the Cuban and Soviet airlifts to Angola and Ethiopia in the 1970s. The rapidity of aircraft ensured that an imprint of force, a tripwire of troops on the ground plus their immediate supply requirements, could be delivered speedily even while the bulk of the presence, plus logistical support, came by sea. In 1950, the first American troops to arrive in South Korea, in response to invasion by North Korea did so by air from Japan, while Britain flew troops and supplies into Amman in 1958 and Kuwait in 1961 when the governments of Jordan and Kuwait feared attack by Arab neighbours. In response to disorder in the Dominican Republic in 1965, the USA airlifted 23,000 troops and supplies in fewer than two weeks.

The airlift of supplies was a classic instance of 'just-in-time' logistics, but the risk to units of supplies being interrupted was an inherent weakness of such logistics as it drove up the risk factor. The military are habituated to chaos by the nature of their systems being able to operate in wartime, and logistics have to be able to operate accordingly, but also not to exacerbate the more general dangers of risk and chaos.

Airlift was both a military enabler and a key element in alliance diplomacy, the latter an important task of logistical capability.[3] Thus, in 1973, in response to the Soviet airlift of weaponry to Egypt and Syria during their Yom Kippur War with Israel, the United States rushed supplies to the Israelis by air, in part to Ben Gurion Airport at Tel Aviv via their air base in the Azores. Initially, the Americans agreed to provide 100 AIM-9 Sidewinder air-to-air missiles, which Israel had ordered before the war, but not the combat aircraft Israel sought while the fighting was continuing. Supplying these aircraft would ensure replacements for Israeli losses, but would compromise American relations with Arab allies, as well as having to be taken from the American military inventory. Israel, however, was under pressure, which, combined with the Soviet airlift and public pressure, led the Americans to provide 40 Phantoms. The Americans failed to find a commercial carrier ready to lease aircraft to transport them, so that USAF Military Airlift Command was ordered to deliver the aircraft directly, rather than, as originally discussed, handing them over to Israeli at Lajes Air Base in the Azores, which would have been less compromising politically. In Operation Nickel Grass, C-141 Starlifters and C-5 Galaxies were used, delivering 22,325 tons of tanks, artillery and ammunition, while with aerial resupply of fuel the Phantoms and Skyhawks flew across the Atlantic ready for combat. In contrast, all the material sent by sea, 33,210 tons worth, arrived only after the war was over.[4]

The reliance of Middle East powers on foreign suppliers was a central element of the politics of logistics, with spares and training important to sustaining these processes and politics. Reliance on spares was shown in May 1967 when the Soviet Union rejected the Egyptian request to provide spare aircraft parts, thus forcing Egypt to abandon plans to launch a pre-emptive attack on Israel. At times, foreign powers would also operate the weapons, as during the Israeli-Egyptian War of Attrition (1968–70) when the Soviets provided SAM anti-aircraft missiles and MIG aircraft, and operated both.

Airports became major sites for both aerial support and operations, as, in the 2000s and 2010s, with Bagram and Kandahar in Afghanistan, Baghdad and Basra in Iraq, and An Nasiriyah near Damascus. To support the International Security Assistance Force (ISAF) deployed in Afghanistan, Kabul International Airport was reactivated and rebuilt in early 2002, permitting the use of C-130 and An-124 aircraft and the operation of an air bridge. From January to July, 38,000 tons of freight, four million litres of fuel, and 28,000 personnel were moved in support of ISAF via the airport.[5] Similarly, the African Union Mission in Somalia, established in 2011, had its main base adjacent to the airport in Mogadishu. Aerial refuelling greatly extended the logistical range of air supply.

Yet scale remained an issue, as with the Soviet intervention in Czechoslovakia in 1968. Tanks were flown into airports, notably Prague, as Antonov-12 and -22 transport aircraft could move them as well as troops and supplies. First flown in 1965, the An-22 could carry a cargo of 80,000 kg (176,350 lbs). At the same time, the Warsaw Pact intervention in Czechoslovakia underlined the significance of ground intervention: the United States at the time could have airlifted no more than 10,000 troops per day from North America to Europe, whereas the Warsaw Pact moved in 250,000 troops within twenty-four hours. The Soviets knew of this weakness. For sustaining operations, sea/land links were seen as more crucial than air ones both during the Cold War and thereafter.

Soviet airlift capacity increased greatly in the 1970s and 1980s with the development of long-range heavy-lift transport aircraft. First flown in 1982, the Antonov An-124 Condor had a wingspan of 73.7 m (240 feet 5 inches), was the world's largest aircraft, could carry a cargo of 150,000 kg (330,700 lbs), and could hold 451 passengers. It held world records for payload, altitude, and distance, and its rugged design made it possible to use a wide range of airfields. Its payload exceeded that of American rivals, notably the C-5 Galaxy, first flown in 1968, which could carry 118,387 kg, and in 1979 delivered a batch of five F-5s to Singapore, and the KC-10 Extender, an aerial refueller which began operating in 1981 and, if configured for cargo, could carry 77,110 kg. The number of Soviet transport aircraft rose to 600 by 1984, although airlift required long and strong runways which added a new form of cost and vulnerability. Moreover, if the cargo cannot be handled and then deployed off-base, the whole system becomes clogged up.[6]

Although both Britain and France developed the idea of a fleet of long-range transport aircraft to reinforce anywhere in the Empire or ex-Empire as needed, other states lacked the capabilities of the USA and the Soviet Union, but most also did not have their requirement for long-distance operations. Instead, there was, by necessity, a fitness-for-purpose situation. Thus, logistics was an aspect of the enhancement of organisation seen in the Chinese Communist army after the Second World War. The Huai Hai campaign in the Chinese Civil War saw the mobilisation of over five million peasants many pushing wheelbarrows full of supplies.[7] So also with the improvement in Chinese logistical capability after intervention in the Korean War in 1950. The Chinese forces sent into Korea in 1950–1 had inadequate logistical support, notably in resupply. Moreover, their system was affected by Allied bombing. As the pace of advance fell and the front line was increasingly consolidated so logistics improved.[8]

Given the significance of China in modern global politics, it is striking how far Chinese logistical tasks were very different to those of the United States. The conflict with India in 1962 was short-term but that with Vietnam from 1979 was longer while there was a major confrontation with the Soviet Union from the 1960s. Yet all of these entailed the need to support commitments solely in frontier areas, while China also had only a modest navy until the 2000s.

Elsewhere in the world, there was logistics that sought to match different physical environment as well as military tasking, notably the length of the conflict and the objectives, especially those of control over territory. This was not an easy process. Thus, in 1948, the 10,000 troops sent from Egypt to fight against the newly-founded state of Israel had no battle experience and, operating with very long supply lines across the Sinai desert, no more than three days' supplies. This ensured that there was no sufficient food to feed the animals, which meant that cavalry had to fight as infantry and limited the delivery of supplies to the troops. The Israeli supply lines, in contrast, were very short, and their logistical system pretty basic, largely a matter of lorries moving in convoys and of a society mobilising nearly its entire resources. Fortified settlements provided supply depots, but also had to be supplied. The logistical chains of the combatants in 1948 were wideranging, both Israel and its Arab opponents drawing on foreign sources for weaponry. With funds raised in the United States, the Israelis purchased weapons there and shipped them to Israel. Arms were also obtained from Czechoslovakia. There was also counter-logistics, as in Italy in 1948 where the Israelis sunk a ship with rifles for Syria and destroyed aircraft intended for Egypt.[9]

Physical and military environments could both be difficult, as in the Biafran War of 1967–70, when the Nigerian army needed to sustain a struggle designed to regain control over an entire separatist area, and in a region that was heavily forested and with few roads. In contrast, in very different conventional conflicts, such as the Arab-Israeli wars of 1948–73, the three Indo-Pakistan wars of 1947–71, and the Somali-Ethiopian conflict over the Ogaden region in 1977–8, lorries were appropriate for fast-moving operations in often flat or semi-arid terrain. Lorries proved the best way to move supporting troops and supplies, although in Chad in 1983 and 1987 the emphasis was on lighter vehicles leading to what was known as the 'Toyota War' with Libya, which itself had civil wars involving similar vehicles from 2011. The dependence on fuel was underlined by the impact of the destruction of infrastructure linked to oil production.

Roads were not only of significance – but also individually more so where they were less frequent – but the use of roads also posed issues. Thus, the Egyptian intervention in 1962–7 in the civil war in mountainous Yemen saw a classic problem of a military suited to a different context. The mobility brought by mechanised vehicles brought an inflexibility and vulnerability in the face of the Royalist technique of ambushing road-bound Egyptian columns. Conversely, in the 1962 war between China and India in the Himalayas, a conflict with front lines unlike that in Yemen, the Chinese benefited, as a result of road building to and along the watershed of the Himalayas, from superior logistics. However, this proved a limited war in which prospects for exploitation were gravely restricted by environmental and logistical factors. In contrast, in 1979, the PLA was hobbled by a lack of roads and navigable rivers running north-south during its unsuccessful intervention in Vietnam, just as earlier had been Mongol, Ming and Manchu China.

At a separate level, logistics were an aspect of the geopolitics of the Cold War, most notably the rivalry between the Eurasian-based Communist bloc and the Western-dominated oceanic sphere. In the Cuban missile crisis of 1962, the naval weakness of the Soviet Union enabled the United States to impose a blockade of Cuba. In part, the logistical requirement for Western powers was for deployment and supply across and from the sea, and, as a result, maritime bases were crucial to the Western powers. American bases in the Pacific, especially the Philippines, supported American operations in East Asia, with Japan a major logistical base in the Korean War (1950–3). Similarly, distant operations posed major logistical requirements for the Royal Navy, the Korean War leading to a significant reliance on Singapore.[10] American participation in the subsequent Vietnam War depended heavily on maritime supplies to the area, followed by land or air deployment, but cargo aircraft were also used, with Hawaii and Okinawa as supporting bases.

Although not part of the Cold War, the British had a similar use of an island base, Ascension, to support the reconquest of the Falkland Islands in 1982. It was used as a base for operations, and the aviation storage facilities controlled by the Americans were made available. Separately, NATO let Britain release stocks of equipment and munitions.

This reconquest indicated the range of modern logistical requirements. The fleet dispatched included 14 Royal Fleet Auxiliary (RFA) tankers, the six Round Table class landing ship logistics (LSL) ships and other RFA ships. Civilian ships, ships taken up from trade (STUFT) were requisitioned or chartered. The ships obtained reflected the needs for range and endurance, but there were also specific requirements such as roll-on/roll-off vessels for vehicles in order to avoid the need for unloading facilities. 15 tankers were requisitioned.

The Argentinians encountered problems with supplies. Merchant navy blockade runners were used and one was sunk by a British helicopter. Three Falkland Islands Company ships were seized by the Argentinians and one ran aground after attack by British ships. Argentinean aircraft had to return to the mainland to refuel after each sortie, which dramatically cut their time in the air near the Falklands.

Meanwhile, aircraft-refuelling capability became part of the British logistical infrastructure. The British used 16 Handley Page Victor aerial refuelling tankers to support air attacks on the Falklands by Vulcan bombers. Their successful attack on Port Stanley airport was intended to prevent aerial resupply of the Argentinian garrison. In addition, cargo aircraft were sent to Ascension Island, where a fuel depot was established to aid the air bridge, while Hercules aircraft fitted with auxiliary fuel tanks were used to airdrop supplies to the fleet. The range of requirements demonstrated both their unpredictability and the need for improvisation.

Notwithstanding the poorly-prepared run up to the conflict, the decision to take military action was made and implemented with immediate effect. The work across the country to prepare for the journey south was an outstanding example of 'can do' and involved people from all walks of life. That it was done so quickly threw up anomalies but the spirit and will was there to sort things out. Supplies was a key instance. Many ships carried stores that were not applicable to the role because it was a case of getting them embarked and sorting out the situation later. On the way to Ascension Island each crew reviewed holdings, re-stored them as necessary and set those not required aside ready for transfer, which resulted in much signal traffic and then a plan for redistribution. Some of the supplies went to the appropriate ship whilst the 'not sure' items were landed ashore where a team had been established to sort things out, and to process stores air freighted. Shortages were also notified and either found from surpluses elsewhere or air freighted from Britain. Some items that were not required were returned to Britain because space was at a premium.

Improvisation was seen in the provision of equipment. Some of the Sea Harriers left Britain with concrete blocks in the nose instead of the radar. Staff in Britain negotiated with the Indian navy to divert their deliveries and these were air-dropped into the sea, picked up by helicopter and then installed. To reduce spares requirements, some routine maintenance periodicities were extended, while, because, in general, the air war near and over the Falkland Islands was prosecuted during daylight hours, night-time was used to work on the aircraft not on standby, notably extending the running hours of the engines beyond the norm. Equipment was kept serviceable to reduce the pressure on the repair teams and the logistical system. Clingfilm from the galley was put over the instrument panels when the aircraft were on manned readiness to fly as the high humidity caused the condensation that was bad for electrical equipment. Black boxes were dried out in the galley ovens, and plastic skin from the Sick Bay used to seal items and cabling vulnerable to water. More generally, under the shock of conflict, particularly losses such as the *Atlantic Conveyor* with the supplies it carried, notably helicopters, it proved necessary to replan and reorganise.[11]

Thus, the Falkland reconquest shows how much the 'fitness for purpose' theme, and the related emphasis on expedients, both seen throughout this book, can be reiterated for the modern, technologically more advanced, period. So also with tank repair. The ability to repair equipment in the field has always been an important aspect of logistical support, notably with the extensive use of wagons. Petrol-driven vehicles added greatly to the complexity. Non-battle losses of tanks through mechanical failure were apt to be greater than battle losses, while overnight repair of equipment and its return to the battle line was highly important to capability. This factor was important in the Second World War, as with the British in Normandy in 1944. In October 1942, the British established the Royal Electrical and Mechanical Engineers, mainly in order to repair tanks and other vehicles on or close to the battlefield. In the Six Day War of 1967, the conflict in Sinai underlined the key role of field maintenance and repair, with the Israelis more effective than the Egyptians in both tasks.

The assessment of logistical capability in this period is in part bound up with those of the strengths respectively of insurrectionary and counter-insurgency warfare. There could be a close relationship between the two, but they were also separate, as in the Vietnam War where American logistical prowess could not make up for strategic failure. Yet, the overlap could be instructive. For example, failure in Vietnam culminated in 1975 with the overthrow of the South Vietnamese army and government. This can be explained in terms of the inevitable superiority of insurgent forces, but the reality was far more complex and logistics played a role. In part, that army made mistaken operational choices (just as the Kuomintang army had done in 1948), and in part it suffered the strategic disaster of having a force configured to be dependent on the American-supplies necessary for its conventional forces, notably spare parts, ammunition and fuel, only to have these requirements rejected by Congress.[12] Conversely, the Greek Communists were defeated in 1949 in part because of the consequences of seeking to pursue a conventional war, but also because the Soviet-Yugoslav split of 1948 compromised their supply system.[13]

Access and bases remained significant to the geopolitics after the Cold War. Saudi Arabia acted as a logistical base for the American-led UN forces committed to the Gulf in 1990–1. This commitment was a major development in logistics for the American armed forces, at strategic and operational levels. Simply moving the powerful American 'heavy' divisions to Saudi Arabia, and then deploying them into the desert before the attack was an incredibly complex achievement. The M1A1 Abrams main battle tank was a major constraint. Aside from its bulk and weight, it has essentially a jet fighter's engine and gets 0.3–0.6 miles a gallon under good conditions. Like other vehicles, it also needs water.

Conversely, in 1999, there was no NATO land invasion of Serb-held Kosovo to supplement the air offensive, as any invasion was dependent on the support of neighbouring countries. Seeking to win Western backing, as well as supportive of the Kosovars, who are fellow Muslims, Albania was willing to do so, but it had very poor transport infrastructure as well as mountainous terrain in the way, and, while Greece was a better fit in both cases, it did not wish to support the Kosovars against Serbia, a choice in which religion and geopolitics both played a role. The logistical issues faced in supporting any deployment strong enough to defeat the Serbs on the ground were formidable.

Similarly, Turkey was unwilling to help with access for a ground attack on Iraq from the north in 2003. In contrast, Pakistan was willing to provide bases and access for the American assault on the Taliban in Afghanistan in 2001 and thereafter provided a key element of the supply route, one made necessary by the inability to rely solely on airborne supplies.

The Americans faced these new tasks, and the related logistical challenges in the unfamiliar physical and political environments they encountered from the 2000s, while also adjusting to a new organisational structure.[14] Divisions were replaced by brigades as part of a streamlining that had implications for logistical structure. After 1991, the USA, Britain and other countries moved to a 'lighter footprint' structure of armed forces for intervention, relying on improved technology and computerised systems. American logistics was characterised by growing scale, complexity, professionalization and specialisation, as well as the continuing application of research on existing systems.[15] The rise of interlinked computers and IT facilitated the rapid analysis of logistical problems and requirements, as well as the use of just-in-time supply carefully linked to the needs of individual units. Employed in the Gulf Wars of 1991 and, even more, 2003, when the Americans deployed fewer troops, this new technology helped make the manoeuvre warfare envisaged by the Americans during the Cold War both more effective, and more effective with the use of fewer units. At the same time, more established practices continued. Thus, in 1991, American KC-10 and KC-135 aerial refuellers conducted about 51,700 refuelling operations, delivering 125 million gallons of fuel. Moreover, logistics have become so much better in the American military that it has become a significant factor in morale. In Iraq in the aftermath of the 2003 invasion, the troops could eat lobster in the mess every evening, to try to make the occupation counter-insurgency more endurable.

The provision of supplies could support counter-insurgency forces, but their presence did not ensure success. For the latter, the counter-insurgency practice of hitting the logistics of insurgent groups remained especially important, which serves as a reminder of continuities in logistical doctrine and practice. A key element was that of preventing local people, whether sympathetic, frightened or for benefit, providing food. This was done successfully by, for example, the Spanish government against anti-Francoist guerrillas in the 1940s and early 1950s, and by the British in Malaya in the early 1950s, for example with the destruction of crops in the village of Jenderam in 1951. Such practices serve to emphasise the overlap of logistics with many other aspects of war.

At the same time, the technological context changes, with new information platforms affecting organisational capability. The Internet was followed by the world wide web, then Web 2.0, and then Social Media. Information about needs can now be transported more rapidly than ever before. More mundanely, logistics were eased by the use of lightweight substances for carrying stores. Thus, in Afghanistan in the 2000s, CamelBak, an American company that, among other products, makes military-grade equipment, came up with a 'Lumber Reservoir Carry Pack', with a feeder tube that could be sucked on while walking, and a cloth outer covering to prevent excessive sweating by the wearer. As throughout, the availability and provision of water will remain a key theme of logistical need and capability.

Chapter 11

Into the Future

Capabilities today provide a guide to the future, and logistics are a key element in these capabilities. Thus, the shortage of British tank transporters is of great consequence for the current rationale of British armour: that it should be part of the NATO defence in Eastern Europe against Russian advance, although the British are now going for lighter armoured vehicles such as the Coyote and Jackal. And so also, aside from tasking, with capabilities in other spheres: in part they are integral to the particular platform, but there are also many systemic issues, notably the reliance on just-in-time practices, and not on traditional store systems with their major requirements for large inventories and space. It is also difficult from a just-in-time mindset to gauge likely maintenance and repair requirements in their scale and timing. Moreover, linked to the issues involved in maintaining existing weaponry, comes the sustaining, repair and construction of transport systems, including in advancing into opposing territory.[1] Thus, combat engineer groups are an important aspect of logistical capability, and will remain so, and it is necessary for them to have the relevant equipment.

Logistical problems are not simply at the margins of military activity, or a matter of states being expected to operate in distant areas for which they lack experience, which may be the case for China. As far as Britain, which has had this experience for centuries, is concerned, there are many problems with its logistical capability, and these problems are of more general relevance and help clarify the question of potential today. Britain is the world's seventh largest economy, but the cost of modern systems leaves them fragile.

Take the key area of power-projection, one in which Britain was the world leader in the periods covered by chapters four to eight. The technology now is far more sophisticated, but digging deep reveals a host of problems. Thus, Britain's warships as a whole appear poorly supplied with the live weapons needed for the execution of their deterrent and combat roles. This appears particularly an issue with AMRAAM air-to-air missiles and other ordnance, with competition between the RAF and the navy. As a result, there are questions about whether ships and RFA (Royal Fleet Auxiliary) magazines will be adequately supplied prior to deployment. The major ships, the *Queen Elizabeth* and the *Prince of Wales* carriers, were seen as crucial to any major distant naval deployment, including, in concert with allies, to East Asian waters where one was proposed for 2021.

However, in 2019, the British carriers had no workshops built into the hangar for the maintenance, tuning and repair of critical aircraft equipment, as reliance was placed on the housing of spare parts and replacement equipment within a container. Without workshop spaces, it is likely that the air group could very quickly have few, if any aircraft, serviceable to fly. More generally, data on serviceability is vital when establishing the size of the F-35B air group to be embarked. Linked to this is the turnabout time between operational missions: in place of the necessary rapid download from the aircraft when still airborne, there are problems with the carriers' current low-speed computer capability, the time taken to download the data taken from each flight, then to send the data ashore for analysis, and then to transmit the intelligence data back to the ship.

Spare engines are an issue due to the size of the F-35's, and there needs to be the capacity to change their engines. The aircraft supply side is highly important with regard to aircraft components that require replacing immediately, ensuring a priority requisition and demand system that needs to be dealt with locally. However, the Automated Logistics Information System (ALIS) used by Britain is a problem for the use of the carriers as the system has hitherto had a bad press. ALIS has control of the global distribution of spare parts required by front-line squadrons, but, if the ALIS module on board fails, this could ground the air group or at best limit sortie generation. The establishment and maintenance of a demand system is crucial for operational effectiveness, and is designed to ensure that when the RAS (supply-ship replenishment at sea) connects with the carriers it provides the necessary supplies, including spares.

The maintenance of stealth capability is also a serious problem. For example, the process of repairing or renovating the stealth coat covering of the F-35 is a very toxic affair and a significant health hazard. Even ashore, with all headquarters squadron facilities available, it take several weeks to do the work properly, but in the salt-ridden maritime environment, the stealth coating deteriorates at a far greater pace than on land, with a serious impact for the penetration of enemy defences. In 2020, the National Audit Office reported that the navy had only one supply ship able to keep the carriers stocked with food and ammunition while on operations. Each carrier cost £3.1 billion, but, at that stage, the airborne radar system crucial to its defence was running eighteen months late, while the full cost of supporting and operating Carrier Strike was unclear as, given the cost, was the ability to meet other naval commitments. The nature of the logistical challenge and cost was seen by the major upgrade work carried out on the jetties at Portsmouth to allow the two British carriers to berth together in their home port.

The complex nature of modern warships ensures that logistics in part depends on base availability. Thus, in the case of the projected increased British deployment to the South China Sea, there was a problem that Singapore, once the key base in the region, would not be able to provide significant logistical support to British forces, bar occasional last-minute breakdown repair as happened to HMS *Argyle* in 2019. Politically, the Singaporean government is very reluctant, such that Changi Naval Base would not be open for sustained logistical support, and the existing British facilities are not adequate for large vessels support. Instead, Britain followed very closely the development of an American presence in Darwin as well as the Japan-Australian Visiting Forces Agreement concluded in 2019. Darwin, Sasebo and Yokosuka, good places to locate, sustain and protect assets under most conditions, were all quietly tested during recent British naval deployments in the region.

Darwin, however, is a classic instance of the trade-offs in logistical bases. It is well-placed to work the Indian Ocean and South China Sea, and far closer than the fleet base in Sydney which demands an almost full week steaming to reach the periphery of South-East Asia. However, the downside is that Darwin lacks a rear and is very isolated past the immediate coastal area. Propinquity in the South China Sea is enjoyed by China, and bases can be seen as launching pads to keep assets operating at the best possible availability and readiness rate.

The argument for large aircraft carriers, American, British and others, is in part logistical in a political context. A modern air force base takes years to set up, and must usually be built in someone else's territory which entails political permission. In contrast, a carrier, although requiring a naval squadron for protection, is a sea-mobile airbase that solves these problems. The US Navy, which in wargames is apt to underrate the risk entailed, treating its carriers as unsinkable, has developed the idea of 'Air Sea Battle,' that, rather than need bases from allies against China, an entire floating and self-contained war machine of ships could sail from the naval base at San Diego and launch a conflict against China, starting from about 1,000 kilometres out, using missiles and airpower.

Points made concerning the American F-35s and the British carriers about cost and practicality can readily be multiplied for other modern weapons platforms, and militaries, such as new British frigates and German submarines. The USS *Fitzgerald* required a $327 million rebuild after colliding with a container ship off Japan in 2017. Commissioned in 1998 and costing about $760 million to build, the USS *Bonhomme Richard*, an amphibious assault ship that in effect was a small carrier, was badly damaged by fire in 2020, although the one million gallons of fuel on board did not ignite. The likely cost of both repair and replacements, and the time entailed, were such that a major loss had occurred.

These issues led to speculation as to how such units might operate in practice, and not least in the face of the strains of symmetrical warfare. These issues also raise the more general questions of the impact of operating cutting-edge militaries, the extent to which technological advances and applications undercut their very value, and certainly use, by their cost, fragility, vulnerabilities, and the problems posed by the skill requirements for modern logistics; and, in conclusion, the general issue of fitness for purpose as raised repeatedly in this book.

Linked to this, sub-optimal purchases of armaments for the sake of domestic manufacturing interests exemplified the grounding of logistics in broader considerations. One of these is the extent to which in many states the military run sections of the economy and indeed government, with this role frequently regarded by the military as a necessary form of political activity. This is logistics in terms of a particular form of professionalization and profit sharing.[2]

Technologies of the future might transform the basic logistical paradigm, the human, including their requirements for support. The characteristics of humans, in the shape of their tolerance to circumstances, such as G-forces, would need to be re-engineered to match even a portion of the possibilities for weapons and delivery systems. Remote-controlled weaponry, whether missiles, drones, or cyber-weaponry, is an alternative, as is the development of robotic warriors or, indeed, genetically-altered ones. They all face issues of supply as well as control, but they certainly offer change, as very much does the capacity of the 'internet of things', and its implications for provisioning and supply.

In the meanwhile, however, there is the continuance, sustaining and, in part, development of what can be seen as legacy systems or, at least, old weapon types that have been upgraded, such as tanks. In the supply of these, air, sea and land transport remain significant, and, in combination, offer a range of possibilities. Aircraft attract the most attention, notably due to their speed. Thus, in 2020, Russian Ilyushin supply aircraft delivered goods to the al-Khadim air base near Banghazi for Marshal Haftar's rebel movement in Libya. Syrian mercenaries were part of the package, not least to secure vital oil installations. This was an aspect of a war that involves rival world powers, and is linked to their provision of supplies: by air, sea, and land. The importance of rapid long-range deployment explains the significance of acquiring modern air transports, notably the Boeing C-17 Globemaster. Aircraft can carry large quantities of weaponry and spares but still remain, and are likely to remain, limited in their capacity to carry volume and weight, certainly when compared to container ships. The contrast between aircraft and shipping is even more striking in the case of transporting fuel, which is very much a bulk product. Thus, if possible, there is a need for the prepositioning of heavy equipment, alongside the rapid deployment of troops by air. Prepositioning, however, depends on bases, and issues with their availability and vulnerability in turn encouraged continued interest in air freight and, separately, commitment to aircraft carriers.

Consideration of the future raises the issue of the contextual interaction of the settings for warfare, within a world with over eight billion people, most of whom live in cities, as well as the tasks, notably maintaining control within states as opposed to fighting other states, and the assumptions and costings bound up in weapons procurement and their intended and real use. These are not issues that vindicate any form of technological triumphalism. Warfare that may be global, including space systems, and yet also very small-scale and local, requires a tasking-based approach to logistics that raises questions about the value of relying on the technologically-focused capabilities with which the chapter began. This tasking will be related to the alliance nature of capabilities, as in the availability of bases and overflying rights and the terms on which they can be used. For example, whereas, America after the Second World War, was able to use the Army Corps of Engineers to build and maintain military facilities, notably air bases, across much of the world, this process became more difficult after the Cold War ended.

In turn, China's bases have become more significant. The Chinese have developed Gwadar in Pakistan, Hambantota in Sri Lanka, and Mombasa in Kenya as they project a naval presence into the Indian Ocean, essentially benefiting from debt traps in order to take over these key, strategic ports. This is linked to a clear policy of naval power-projection, with a blue water fleet as both means and expression. In part, this power-projection serves the logistics of China's strength by safeguarding investments and resource-flows, notably of oil from the Middle East.

There is also separately the question of the 'social bargain' or, rather, interests and shared interest, both international and domestic, that underpin logistical systems and provide logistical resources, and why, how, and when, those will change. Again, context and task are to the fore. As is clear throughout the book, a lot will depend on how far this support is related to conflict within states and how far between them; and, again, how far the organisation in question was the state and how far non-state organisations. This, again, raises the question of what composes the state. In many senses, that is a matter raised throughout any discussion of logistics and one identified as significant by the 'stadial' writers of the eighteenth century such as Adam Smith.

Looking forward, it is difficult to see what consent will mean in terms of major states seeking to deploy large resources for the sake of major logistical commitments, and notably so if abroad and not to be raised by means of enforced contributions. There will probably be a parallel with French war-making under Louis XIV (r. 1643–1715), in that there will be resources for the opening campaign, but problems thereafter as weapon supplies are depleted. Just-in-time techniques, including 3D production, will address some of these, but there will be problems in terms of meeting the actual resource cost. As part of this question, the reliance on credit is hardwired into modern states, but it is unclear whether that has been exhausted by the extent to which Quantitative Easing since 2008 has led to a substantial fall in interest rates. The willingness of international investors to offer credit on these terms is unclear, but there are few alternatives to interest-bearing government debt.

As a consequence of financial and political imponderables, it will not be possible to anticipate the resilience of the supply chain, a resilience ultimately dependent on credit at a number of scales. This difficulty should be a major restriction on any prediction for logistics once we move from discussing the machinery used. For example, public-private partnerships ultimately rest on the willingness of private concerns to co-operate, or on the capacity to coerce them. Neither, the first certainly in the absence of a return, is a happy prospect, which raises questions about the viability of war that are separate to those of the limitations of particular weapons systems, limitations that in part arise from the limited numbers that arise from their high-cost specifications. Tensions in the terms of public-private partnerships were seen in 2019–20 in rifts within the élites in Syria and Saudi Arabia as their governments, both in economic difficulties, sought to sustain wars and, to do so, extorted money from figures close to the regime.

Another sphere of limitations is that of the very use of force in order to obtain outcomes (as opposed to output), and that in a world of unprecedented numbers of people, as well as strongly-held ideologies. As warfare will continue, the question then arises as to whether, in a more populous future, there will be a greater role for the private provision of violence, maybe by delegation from the state, but also in despite or, indeed, defiance of it. This would be a revival of earlier practices[3] and thus a reminder of the degree to which a developmental model of the subject, with the past as somehow obsolete and irrelevant, is flawed. At the same time, the question of revival throws to the fore the issue of how far there has been significant, even paradigmatic, change as a consequence of new technology, tasking and setting, and how far this process will continue. Separately, we have the issue of the survival of evidence which is much worse for non-state actors, thus contributing to a misleading account of conflict, capability and logistics.

Given the factors already cited, the character of logistics in part is a matter of how force, conflict and the military are defined – all issues that in part are contextual. If maintaining civil order against violent opposition is a factor, then police and paramilitary forces are as much part of the logistical chain as regular armed units, a point that is underlined by the protected nature of their respective bases and by their weaponry. The balance between offensive and defensive logistical capabilities is affected by the inclusion of such forces.

That point overlaps with the weight to be placed on the domestic (internal) use of force, and, seperately, the question of the timing of action, which will remain a factor. For example, the Anglo-American practice in the Second World War, a 'fire-power, support-heavy way of war', took longer to organise,[4] but subsequent use of smaller and lighter forces, combined with reliance on air transport and more precise air support speeded up this process, as the Americans showed in the Gulf Wars of 1991 and 2003. However, aside from the degree to which victory over conventional forces does not necessarily provide the expected outcome, there is the broader point that conflict might take longer.

Moreover, as asymmetrical warfare demonstrates, militaries face the problem that deploying and using their strength may be too slow, at the tactical, operational and indeed strategic levels to confront lighter opponents, and notably so if the latter lack clear bases and an obvious military structure. Counter-insurgency is not technology-light, but it is technology-different and logistics-different.[5] That this is not the sole type of conflict underlines the need for multi-purpose capabilities, and for doctrine accordingly. That this debate has occurred already, does not make it less relevant today.[6]

Separately, the highly-specialised nature of modern logistical support increases the logistical burden, and thus the relative disadvantage of this support system. The costs, vulnerabilities, and limited applicability are all issues, and cut across established principles of logistical effectiveness, notably flexibility, durability, and economy of effort.

Alongside these constants will come variations reflecting the very different nature of modern societies. The prime causes for these differences are political, social and economic, and not technological. The higher cost of military labour and professionalism (including the training and social welfare costs) in some societies, notably the United States, encourages the use of less expensive contractors for many logistical tasks. Separately, the dependence there of the military on private companies for the research and production of weaponry provides a particular character to the 'military industrial' complex. This situation entails drawing on the resources of the economy in a manner that is different to that in countries where circumstances, need and ideology produce a contrasting outcome.

In China, although there is now a significant private sector, it is far more under the control of the government than in the case of the United States. Separately, the effectiveness of Chinese logistics was rapidly demonstrated in the summer of 2020 when, alongside confrontation with India over the contested border in the Galwan Valley, China installed a new road and camp there, as well as deploying the PCL-181, an effective self-propelled lightweight gun to nearby Tibet, using the recently expanded rail system. The global range of Chinese logistical capability is increasingly an issue, but reading from one aspect or circumstance to another, which is so often done in military history, has to be handled with care. Moreover, global range brings an emphasis on air and sea transport very different to land operations in border areas. As ever, logistics is set in large part by the specific tasking.

A part of the future that recognisably continues current trends is climate change. This could go into reverse, but there are no immediate signs that it will, and, even if it decelerates or holds constant, it is still a highly significant development in the environmental context for human society as a whole, for conflict, and for logistics. Contextual changes include the opening up of sea passages to the north of Asia and North America and the increased strain of operating in areas of the world affected by heat and water shortage. If the basic workings of the global economic order and network were to be affected by serious climate change, that would be both a major possible trigger for armed conflict and a cause of major strain for logistical systems. More frequent large-scale storms will affect maritime and air navigation, while a rise in the sea level will make military operations in littoral areas more difficult. Control over basic food supplies in a situation of shortage may make military control harder or more problematic.

Consideration of the impact of climate change emphasises the basic theme of the book that logistics is inextricable from contexts, notably environmental, military and political ones. Such consideration also brings back the issue of food availability to full prominence, in comparison downplaying that of fuel, which was to the fore for the twentieth century. Military and logistical planning will have to consider the impact of climate change, and to do so at the strategic, operational and tactical levels.

Chapter 12

Conclusions

'If we follow an army on its progress, desolation and depopulation goes hand in hand. Carrying off everything of value, men, women, and children included, and burning what they cannot carry, they render the country they have overrun untenable on their retreat. Those who have retired to the fastnesses appear half to die or emigrate from want of sustenance, so that one country leaves the other country overrun and wild.'[1]

Congo in 1997? Syria in 2020? Actually Burma (Myanmar)-Siam (Thailand) in 1785, a reminder, alongside important changes in technology, which was a major theme from the Age of Steam in chapter six onwards, and in other factors, of the timeless character of much of logistics. The crucial dependence on logistics on the natural environment was captured for this very area by Field Marshal Sir William Slim when contrasting British and Japanese performance during the Japanese conquest of Burma in 1942:

The Japanese were obviously able to move … through jungle that we had regarded as impenetrable…. They travelled lighter than we did and lived much more off the country. Nearly all our transport was mechanical, and this stretched our columns for miles along a single road through the jungle, vulnerable everywhere from air and ground…. this being tied to a road proved our undoing. It made us fight on a narrow front, while the enemy, moving wide through the jungle, encircled us and placed a force behind us across the only road. The Japanese had developed the art of the road-block to perfection; we seemed to have no answer to it. If we stood and fought where we were, unless the road was reopened, we starved. So invariably we had turned back to clear the road-block, breaking through it usually at the cost of vehicles, and in any case making another withdrawal.[2]

Similarly, in 1924, Rudolph, 10th Earl of Cavan, the Chief of the (British) Imperial General Staff from 1922 to 1926, noted the need for British forces to adapt to what was there in the shape of the particular environment, which was a major task after being configured for the First World War: 'So much of our work has to be done in uncivilised parts of the world that it is a mistake to pin faith to a type of unit largely dependent on civilised conditions, good roads and open country'.[3]

For war, as for other aspects of life, environmental constraints operate alongside the human attempt to adapt to them and also to overcome them. Logistics is a key element of this process, one of aspiration, method and organisation of matching ends, means and contexts. In 1927, the focus was on logistics when Sir Frederick Barton Maurice, a former Major-General, gave his inaugural lecture as the Professor of Military Studies at King's College, London:

> History shows that great changes in the character of warfare are normally brought about by other forces than the power of weapons … for the tendency is that sooner or later an antidote is found for each new form of attack … the deadlock of trench warfare of 1915–1918 was a revolution, which changed the character of war. But the prime cause of that change was only partly weapons, it was still more numbers; and the reason why armies of millions could be maintained in the field was, I think, first the development of railways, roads, and mechanised transport of all kinds, which enabled supplies to be brought to the front in almost unlimited quantities, and secondly the progress of medical science, which has almost eliminated the danger of epidemic disease … With certain reservations as regards the sea, I would say that the changes which affect the daily lives of peoples, such as developments of transport and of communication, tend to affect war much more than do changes of weapons.[4]

At the same time, logistics were affected by changes in weaponry. Thus, maintenance, and the logistical issues involved, were altered by mass production. Indeed, much of the importance of the introduction of single-shot breech-loaders in the mid-nineteenth century, followed by that of repeating firearms, stemmed from the ability to mass-produce rifling, sliding bolts, magazine springs, and chain-feeds to a high standard, and also to provide the large quantities of ammunition required were these weapons to be used, a need further complicated by the ability of troops to fire all their ammunition away in a matter of minutes. Effective mass-production provided a major advantage over craft-manufactured firearms, however good the latter were on an individual basis.

Mass production, of both weapons and other goods, such as packaged food, had implications in terms of maintenance requirements and also transport, as the accompanying unitisation made it easier to quantify mass-transport logistical needs and responses. As such, logistics could more readily be built into the planning, at once seeking precision, and scale, of militaries from the late nineteenth century. They sought to define and manage risk based upon size, space, time and technology; and logistics was a key element of the related process,[5] although still generally attracting insufficient attention compared to other aspects of planning. Logistics also overlapped with the preparedness of societies as a whole for conflict, a preparedness, for example of food supplies,[6] that can be discussed in terms of total war.

Repeatedly, alongside the need to respond to change comes the value of established methods; with, as a reminder of the significance of vocabulary in triggering reader responses, the possibility of replacing established by traditional or unchanging, each of which produce a critical reception. Thus, the British Ordnance Office faced major problems in responding to the very great demands when Britain unexpectedly began lengthy conflicts in 1739, 1754 and 1775, with Spain, France and the American Patriots respectively; and each with very different requirements. More generally, greatly fluctuating demands for gunpowder created problems in Britain (as elsewhere) for both the Ordnance Office and gunpowder makers, while, whether in government or in opposition, politicians were unwilling to extend the power of government in the crucial sphere of gunpowder production and distribution, even though they were aware of their inadequacy. In general, the established methods helped produce a generally effective response,[7] in part because, as ever, the optimal effectiveness of the system was affected both by the goals that were pursued and by the parameters of the politically possible resources, or, rather, obtainable resources, were, and remain, a key element of this interplay of goals and political possibility.

The timeless character of the resource dimension is captured by the ongoing impact of inadequate finance on logistics and the consequent need for political support. Thus, in 1639, in the First Bishops' War with Scotland, the poorly-prepared English army had its logistics wrecked by inadequate finance, but there was no political cohesion behind raising the necessary funds, a factor that helped to lead to the outbreak of the English Civil War in 1642. This element both encouraged attack on the opposing logistical system, and in the widest sense. Thus, in January 1945, Montgomery wrote of the leading German industrial region: 'the main objective of the Allies on the western front is the Ruhr: if we can cut it off from the rest of Germany the enemy capacity to continue the struggle must gradually peter out'.[8]

There was also the hope of technological leapfrogging in order to overcome constraints. Thus, in an article on the 'Future of War' in *Newsweek* on 25 December 1944, J.F.C. Fuller wrote: 'Before the present century has run its course, there is nothing fantastic in suggesting that complete armies will be whisked through pure speed a thousand miles above the Earth's surface, to speed at 10,000 miles an hour toward their enemy'. More prosaically, visiting India in 1926 on behalf of the War Office, Fuller had stressed the need for (Britain's) Indian Army to mechanise, arguing that the navy and air force 'were mechanised forces, materially highly progressive' while he thought the army there reflected 'its surroundings in being oriental',[9] a process and term that amounted to condemnation.

The types of force available and deployed, and their tasks, were crucial in determining logistical requirements and capability. Thus, with research today: the use of smart weaponry is designed not only to provide the effectiveness of greater precision, but also to reduce the logistical drag. For example, the American Army Research Laboratory's Aeromechanics and Flight Control Group has probed the potential of what it termed the Collaborative Cooperative Engagement Programme. This involved moving the world's leading army from inexpensive 'dumb' weapons to more efficient ones. The plan rests on guiding the 'dumb' weapons by means of radio messages from smart munitions. As a result, a swarm of submunitions would be given a guidance system, replacing indiscriminate fire. 3D printers, including metal printers, also challenge existing economies of scale, and offer the chance for rapid response-led logistical provision.

In conclusion, it is appropriate to reconsider the argument of van Creveld. After his appropriate call for attention to the subject, he suggested that eighteenth-century armies 'managed to do very much better in relation to the theoretical limits of the means at their disposal than do modern forces' because friction increases with complexity:

> Hence, even if an entirely new and unprecedentedly effective means of transport were to appear tomorrow, it is to be expected that only a small fraction of its maximum theoretical capacity will ever be utilised in practice, and that its effect on the speed of mobile operations will therefore be marginal.[10]

Indeed, the book closed on a note of pessimism about the possibility of understanding war and planning for its challenges.[11] That is not the conclusion here. Certainly, high specification warfare has become very expensive, but the logistical burden has fallen due to the use of smaller militaries, and the development of unmanned vehicles. Moreover, the comment quoted above did not prefigure the possibilities of more intensive aerial supply and resupply. Van Creveld commented on the logistical drawbacks of the long-planned invasion of France in 1944, notably the failure to 'allow sufficiently for the inevitable friction of war'.[12] However, invasions more recently, admittedly against easier targets, notably Afghanistan in 1979, Panama in 1989 and Iraq in 1991 and 2003, have delivered forces more nimbly and achieved at least their occupation targets to time. Subsequent problems were strategic rather than logistical. Operations Research, the computerised analysis of the greater flow of information, and both techniques and doctrine taken from civilian logistics, have all enhanced potential. Network-centric warfare linked to the relevant logistical support is possible, although far-from-easy.

Indeed, a key thread in the argument in this book is that logistical capacity is not simply bureaucratic capacity or even state capacity, and/or dependent on technological proficiency. In place of a Weberian tendency to see the co-evolution of the state and warfare as a story of increasing complexity and capability, albeit at substantial economic and human cost, what really matters is the fitness of a means of supplying and conveying an armed force with reference to its particular environment and situation. In many settings, complexity and technological sophistication are not advantages, either because of local ecology, the carrying capacity of the society in question, or because of the very great material costs that they impose with diminishing operational returns.

Moreover, the importance of logistics is situational, such that the same capacities and techniques that might be excellent in one situation, may fail in another. Thus, the speed, improvisation and flexibility that worked for the Japanese in 1941–2 largely failed in 1944–5 with many dying of disease and hunger. Imperial powers could follow different logistical patterns on distant frontiers to those seen in the metropole, the British in India meshing native elements with European organisational input. Logistical requirements could vary. In the Falklands War in 1982, the British found that the standard army boot was not good enough for the conditions faced; a problem, however, that was different to Gettysburg in 1863 when the battle was begun by Confederate troops moving into the town in search of footwear.

Separately, while logistics is usually discussed at the operational level, indeed as an aspect of the conflation of organisational capability and operational art,[13] there is the need to consider logistics across the spectrum from tactics to strategy, with the latter as particularly significant.

Turning away from leading militaries, it is apparent that others have been able to deploy and sustain forces, not least over the last two decades, in civil wars in Congo, the Central African Republic, Sudan, South Sudan and Libya, and also in state-to-state conflicts, as between Eritrea and Ethiopia. The outcome pattern in each of these cases, and in many others, is generally not that of the rapid resolution extolled by van Creveld and sought in the conventional Western model. However, that pattern has never been applicable for counter-insurgency conflicts or, indeed, defensive warfare, and also underplays the extent to which fighting, indeed war, might be seen not as an unnatural state and undesirable pathology separate to peace, but rather as a frequent, even semi-permanent, situation. This ensures logistical commitments that are different in duration and scale to those of the offensive, high-spectrum warfare considered by van Creveld and most writers on the topic. Whether or not insurgency and counter-insurgency warfare, or maybe simply large-scale lawlessness and the attempt to contain it, will be the prime form of conflict in the more crowded and resource-pressured urban spaces of the future, logistics needs to be rethought on a wider canvas. Hopefully this book is a contribution to that rethinking.

Notes

Introduction

1. *Logistics* (1996), p. ix.
2. Jomini, *Précis de l'Art de la Guerre* (Paris, 1838), pp. 140–77.
3. C. Reardon, *With a Sword in One Hand and Jomini in the Other* (Chapel Hill, NC., 2012).
4. G. Perjés, 'Army Provisioning, Logistics and Strategy in the Second Half of the Seventeenth Century', *Acta Historica Academiae Scientarium Hungaricae*, 16, no. 1/2 (1970), p. 3.
5. *Ibid.*, pp. 36–8.
6. J. Fynn-Paul (ed.), *War, Entrepreneurs and the State in Europe and the Mediterranean, 1300–1800* (Leiden, 2014).
7. C. Tilly, *Coercion, Capital, and European States, AD 990–1990* (Oxford, 1990), and 'States, State Transformation, and War', in J.H. Bentley (ed.), *The Oxford Handbook of World History* (Oxford, 2011), pp. 190–1. For Tilly's other relevant arguments as 'too simple', J.A. Goldstone, 'Political Trajectories Compared', in Bentley, S. Subrahmanyam and M.E. Wiesner-Hanks (eds), *The Cambridge World History. VI. The Construction of a Global World, 14000–1800 CE. Part 1: Foundations* (Cambridge, 2015), p. 453 and S. Morillo, 'The Sword of Justice: War and State Formation in Comparative Perspective', *Journal of Medieval Military History*, 4 (2006), pp. 1–17.
8. D. Stoker, F.C. Schneid and H.D. Blanton (eds), *Conscription in the Napoleonic Era: A Revolution in Military Affairs?* (Abingdon, 2009).
9. S. Morillo, 'Mercenaries, Mamluks and Militia. Towards a cross-cultural typology of military service', in J. France (ed.), *Mercenaries and Paid Men. The Mercenary Identity in the Middle Ages* (Leiden, 2008), p. 248.
10. P.W. Singer, *Corporate Warriors: The Rise of the Privatised Military Industry* (Ithaca, NY, 2003).
11. D.M. Peers, 'Revolution, Evolution, or Devolution: The Military and the Making of Colonial India', in W.E. Lee (ed.), *Empires and Indigenes: Intercultural Alliance, Imperial Expansion, and Warfare in the Early Modern World* (New York, 2011), p. 8.
12. S. Subrahmanyan, 'Profiles in Transition: Of Adventurers and Administrators in South India, 1750–1810', *Indian Economic and Social History Review*, 39 (2002), pp. 197–232.
13. J. Glete, *War and the State in Early Modern Europe: Spain, the Dutch Republic and Sweden as Fiscal-Military States, 1500–1600* (2002).

14. J. Brewer, *The Sinews of Power: War, Money, and the English State, 1688–1783* (1989); J.A.F. de Jongste and A.J. Veenendaal (eds), *Anthonie Heinsius and the Dutch Republic, 1688–1720: Politics, War and Finance* (The Hague, 2002); G. Storrs (ed.), *The Fiscal-Military State in Eighteenth-Century Europe* (Farnham, 2009).

15. J. Black, *Geopolitics and the Quest for Dominance* (Bloomington, Ind., 2015), and *Military Strategy: A Global History* (New Haven, Conn., 2020).

16. G. Satterfield, *Princes, Posts and Partisans: The Army of Louis XIV and Partisan Warfare in the Netherlands, 1673–1678* (Leiden, 2003).

17. C.J. Rogers, 'By Fire and Sword: Bellum Hostile and "Civilians" in the Hundred Years War', in M. Grimsley and Rogers (eds), *Civilians in the Path of War* (Lincoln, NB, 2002), pp. 33–78.

18. J. Black, *War and the Cultural Turn* (Cambridge, 2012).

19. J. Brauer and H. van Tuyll, *Castles, Battles, and Bombs: How Economics Explains Military History* (Chicago, Ill., 2007).

20. E. Wald, *Vice in the Barracks: Medicine, the Military and the Making of Colonial India, 1780–1868* (Basingstoke, 2014).

21. P.K. O'Brien, 'A Global Perspective for the Comprehension of Fiscal State Formation from the Rise of Venice to the Opium War', in R. Harding and S.S. Ferri (eds), *The Contractor State and its Implications, 1659–1815* (Gran Canaria, 2012), pp. 233–4.

22. D. Parrott, *The Business of War: Military Enterprise and Military Revolution in Early Modern Europe* (Cambridge, 2012); G. Rowlands, *The Dynastic State and the Army under Louis XIV: Royal Service and Private Interest, 1661–1701* (Cambridge, 2002).

23. W. Cobbett, *Parliamentary History*, XVII (1813), cols 500–1.

24. P.C. Perdue, *China Marches West: The Qing Conquest of Central Eurasia* (Cambridge, Mass., 2005).

25. J. Pryor, *Geography, Technology and War: Studies in Maritime History of the Mediterranean, 649–1571* (Cambridge, 1988).

26. R. Murphey, 'Ottoman Expansion, 1451–1556', in G. Mortimer (ed.), *Early Modern Military History, 1450–1815* (Basingstoke, 2004), p. 52.

27. F.M. Kert, *Privateering: Patriots and Profits in the War of 1812* (Baltimore, MD, 2015).

28. M. van Creveld, *Supplying War. Logistics from Wallenstein to Patton* (Cambridge, 1977).

29. H. Boog et al., *Die Bedeutung der Logistik für die militärische Führung von der Antike bis in die neuste Zeit* (Herford, 1986); J.A. Lynn (ed.), *Feeding Mars: Logistics in Western Warfare From The Middle Ages to the Present* (Boulder, Col., 1993).

30. See, in particular, J. Haldon (ed.), *General Issues in the Study of Medieval Logistics: Sources, Problems and Methodologies* (Leiden, 2006); J. Pryor (ed.), *Logistics of Warfare in the Age of the Crusades* (Aldershot, 2006); J. Haldon, V. Gaffney, G. Theodoropoulos and P. Murgatroyd, 'Marching across Anatolia: Medieval Logistics and Modeling the Mantzikert Campaign', *Dumbarton Oaks Papers*, 65/66 (2011–12), pp. 209–35; B.S. and D.S. Bachrach, *Warfare in Medieval*

Europe, c.400–c.1453 (Abingdon, 2017), pp. 154–212; M. Larnach, 'The Battle of the Gates of Trajan, 986: A Reassessment', *JMH*, 84 (2020), pp. 18–20. For this method, see also D.W. Engels, *Alexander The Great and the Logistics of the Macedonian Army* (Berkeley, Calif., 1978); B.S. Bachrach, 'The Barbarian Hordes That Never Were', *JMH*, 74 (2010), pp. 903–4; J. Linn, 'Attila's Appetite: The Logistics of Attila the Hun's Invasion of Italy in 452', *JMH*, 83 (2019), pp. 325–46.

31. C. Tyerman, *How to Plan a Crusade: Reason and Religious War in the Middle Ages* (2015).

32. J.W.I. Lee, *A Greek Army on the March: Soldiers and Survival in Xenophon's 'Anabasis'* (Cambridge, 2008).

33. M.E. Latham, *Modernisation as Ideology: American Social Science and 'National Building' in the Kennedy Era* (Chapel Hill, NC., 2000).

34. G. Parker, 'In Defense of *The Military Revolution*', in C.J. Rogers (ed.), *The Military Revolution Debate* (Boulder, Col., 1995), p. 355. For a liberal progressivism at Christ's College, Cambridge, on the part of Jack Plumb, Parker's teacher, to whom Parker referred as having 'such an immense impact', D. Cannadine, 'Historians in the "Liberal Hour": Lawrence Stone and J.H. Plumb Re-Visited', *Historical Research*, 75 (2002), pp. 315–54 and 'John Harold Plumb 1911–2001', in 'Biographical Memoirs of Fellows III', *Proceedings of the British Academy*, 124 (2004), pp. 269–309 at pp. 286–96.

35. C.R. Shrader, *History, of Operational Research in the United States Army* (3 vols, Washington, 2006–9); M. Elliott, *RAND in Southeast Asia: A history of the Vietnam War Era* (Santa Monica, Calif., 2010).

36. T. Farrell, *Unwinnable: Britain's War in Afghanistan, 2001–2014* (2017).

37. H. Spruyt, *The Sovereign State and its Competitors: An Analysis of Systems Change* (Princeton, NJ, 1994); E.T. Penson, *The Theory of the Growth of the Firm* (3rd edn, Oxford, 1995); F.C. Lane, *Profits from Power: Readings in Protection Rent and Violence-controlling Enterprises* (Albany, NY, 1979).

38. J.S. Nolan, 'The Militarization of the Elizabethan State', *JMH*, 58 (1994), pp. 418–19.

39. R.M. Eaton and P.B. Wagoner, 'Warfare on the Deccan Plateau, 1450–1600: A Military Revolution in Early Modern India?', *Journal of World History*, 25 (2014), p. 50.

40. B. Teschke, 'Revisiting the "War-Makes-States" Thesis: War, Taxation and Social Property Relations in Early Modern Europe', in O. Asbach and P. Schröder (eds), *War, the State and International Law in Seventeenth-Century Europe* (Farnham, 2010), p. 58.

41. L. Loreto, *Per la storia militare del mondo antico* (2006), pp. 97–107.

42. J.M. Shaw, *The Cambodian Campaign: The 1970 Offensive and America's Vietnam War* (Lawrence, KS, 2005).

43. Memorandum on Anglo-Portuguese meeting, 16 Sept. 1705, BL. Add. 61122 fol. 64.

44. *The Times*, 20 July 2020.

45. J.P. Roth, *The Logistics of the Roman Army at War, 264 BC – AD 235* (Leiden, 1999), p. 332.

46. G. Macola, *The Gun in Central Africa: a History of Technology and Politics* (Athens, Oh., 2016).
47. R. Entenmann (ed.), 'Andreas Ly on the First Jinchuan War in Western Sichuan, 1747–1749', *Sino-Western Cultural Relations Journal*, 19 (1997), pp. 11, 15.
48. See the work coming out from 'The European Fiscal-Military System 1530–1870' project, including access through the website, https://fiscalmilitary.history.ox.ac.uk/home, to P. Wilson 2017 inaugural lecture, 'Competition through Cooperation: The European Fiscal-Military System'; P. Wilson, 'Foreign military labour in Europe's transition to modernity', *European Review of History*, 27 (2020), pp. 12–32.
49. NA. SP. 84/202 fol. 128.
50. S.P. Reyna, 'The Force of Two Logics: Predatory and Capital Accumulation in the Making of the Great Leviathan, 1415–1763', in S.P. Reyna and R.E. Downs (eds), *Deadly Developments: Capitalism, States and War* (Amsterdam, 1999), pp. 23–68.
51. J. Black, *War in Europe: 1450 to the Present* (2016).
52. J. Maiolo, *Cry Havoc: How the Arms Race Drove the World to War 1931–1941* (New York, 2010).
53. K. Helleiner, 'The Vital Revolution Reconsidered', *Canadian Journal of Economics and Political Science*, 23 (1957), p. 1.
54. For a controversial instance, Y. Pasher, *Holocaust vs Wehrmacht: How Hitler's 'Final Solution' Undermined the German War Effort* (Lawrence, KS, 2015).
55. D.S. and B.S. Bachrach, 'Bruno of Merseburg's *Saxon War*: A Study in Eleventh-Century German Military History', *JMH*, 81 (2017), pp. 359–60.

Chapter 1

1. K.F. Otterbein, *How War Began* (College Station, Tx., 2004).
2. A.S. Kolberg, 'There is Power in a Cohort: Development of Warfare in Iron Age to Early Medieval Scandinavia', *JMH*, 83 (2019), p. 29.
3. P. Varley, 'Warfare in Japan 1467–1600', and J. Gommans, 'Warhorse and gunpowder in India *c*.1000–1850', in J. Black (ed.), *War in the Early Modern World* (London, 1999), pp. 61, 63–4, 73 & 107.
4. D.J. Lonsdale, 'Alexander the Great and the Art of Adaptation', *JMH*, 77 (2013), p. 823.
5. A.J. Spalinger, *War in Ancient Egypt: The New Kingdom* (Oxford, 2005).
6. D.W. Engels, *Alexander the Great and the Logistics of the Macedonian Army* (Berkeley, Calif., 1978); A. Chaniotis, *War in the Hellenistic World* (Oxford, 2005).
7. J.A. Olsen and M. van Creveld (eds), *The Evolution of Operational Art from Napoleon to the Present* (Oxford, 2011).
8. J.E. Lendon, *Song of Wrath: the Peloponnesian War Begins* (New York, 2010).
9. J.P. Roth, *The Logistics of the Roman Army at War, 264 BC – AD 235* (Leiden, 1999); P. Erdkamp, *Hunger and the Sword: Warfare and Food Supply in the Roman Republican Wars, 264–30 BC* (Leiden, 1998) and 'Supplying Armies in the Iberian Peninsula during the Republic', in C. Carreras and R. Morais (eds), *The Western Roman Atlantic Façade* (Oxford, 2010), pp. 119–190.

10. S. Stallibrass and R. Thomas (eds), *The Archaeology of Production and Supply in NW Europe* (Oxford, 2008).

11. X. Li, *Bronze Weapons of the Qin Terracotta Warriors. Standardisation, craft specialisation and labour organisation* (Oxford, 2020).

12. From here on, all dates are CE.

13. S. Morillo, 'Battle Seeking: The Contexts and Limits of Vegetian Strategy', in *Journal of Medieval Military History*, 1 (2002), pp. 21–41.

14. R. Higham, *Making Anglo-Saxon Devon* (Exeter, 2008); R. Lavelle, *Alfred's Wars: Sources and Interpretations of Anglo-Saxon Warfare in the Viking Age* (Woodbridge, 2010).

15. R. Abels and S. Morillo, 'A Lying Legacy? A Preliminary Discussion of Images of Antiquity and Altered Reality in Medieval Military History', *Journal of Medieval Military History*, 3 (2005), pp. 1–13.

16. B.S. Bachrach, *Early Carolingian Warfare* (2001) and 'Logistics in Pre-Crusade Europe', in Lynn (ed.), *Feeding Mars*, pp. 57–78. For England, J. Peddie, *Alfred the Good Soldier: His Life and Campaigns* (1989); D. Hill and A. Rumble (eds), *The Defence of Wessex: the Burghal Hidage and Anglo-Saxon Fortifications* (Manchester, 1996); R. Abels, *Alfred the Great* (1998).

17. D.S. Bachrach, 'Early Ottonian Warfare: The Perspective from Corvey', *JMH*, 75 (2011), pp. 393–409 and *Warfare in Tenth-Century Germany* (2012).

18. D.S. Bachrach, 'The Military Organisation of Ottonian Germany, c.900–1018: The Views of Bishop Thietmar of Merseburg', *JMH*, 72 (2008), p. 108.

19. M. Jones, 'The Logistics of the Anglo-Saxon Invasions', in D.M. Masterson (ed.), *Naval History: The Sixth Symposium of the United States Naval Academy* (Wilmington, DE, 1987), pp. 62–9.

20. T.S. Burns, *Rome and the Barbarians, 100 BC-AD 400* (Baltimore, MD, 2003).

21. M. Lower, *The Barons' Crusade: A Call to Arms and Its Consequences* (Philadelphia, Penn., 2005).

22. J.H. Pryor, *Geography, Technology, and War: Studies in the Maritime History of the Mediterranean* (Cambridge, 1988) and (ed.), *Logistics of Warfare in the Age of the Crusades* (Aldershot, 2006).

23. B.S. Bachrach, 'Charlemagne and the Carolingian General Staff', *JMH*, 66 (2002), pp. 338–45.

24. L.W. Marvin, *The Occitan War* (Cambridge, 2008).

25. C.J. Rogers, *War Cruel and Sharp: English Strategy under Edward III, 1327–1360* (Woodbridge, 2000).

26. Y.N. Harari, 'Strategy and Supply in Fourteenth-Century Western European Invasion Campaigns', *JMH*, 64 (2000), pp. 297–334, esp. p. 333. For failure, I. Krug, 'Food, Famine and Edward II's Military Failures', *Journal of Medieval Military History*, 16 (2018).

27. J. France, *Western Warfare in the Age of the Crusades 1000–1300* (1999).

28. H.J. Hewitt, *The Organisation of War under Edward III* (Manchester, 1966).

29. M. Prestwich, *Armies and Warfare in the Middle Ages: The English Experience* (New Haven, Conn., 1996).

30. J.F. Shear, 'Hannibal's Mules: The Logistical Limitations of Hannibal's Army and the Battle of Cannae, 216 B.C.', *Historia*, 45 (1996), pp. 159–87.

31. J. Haldon, *The State and the Tributary Mode of Production* (1993); M. Bartusis, *Land and Privilege in Byzantium: The Institution of Pronoia* (Cambridge, 2012).
32. G. Cushway, *Edward III and the War at Sea: The English Navy, 1327–1377* (Woodbridge, 2011).
33. D.J. Kagay, 'The Defense of the Crown of Aragon during the War of the Two Pedros, 1356–1366', *JMH*, 71 (2007), p. 19.
34. J. Haldon, 'Roads and Communications in the Byzantine Empire', in Haldon (ed.), *Logistics of Warfare in the Age of the Crusades* (Farnham, 2006), p. 158; J. Linn, 'Attila's Appetite: The Logistics of Attila the Hun's Invasion of Italy in 452', *JMH*, 83 (2019), pp. 330, 345–6.

Chapter 2

1. G. Parker, *The Army of Flanders and the Spanish Road, 1567–1659: The Logistics of Spanish Victory and Defeat in the Low Countries' Wars* (Cambridge, 1972).
2. C. Finkel, *The Administration of Warfare: Ottoman Campaigns in Hungary, 1593–1606* (Vienna, 1988).
3. I.A.A. Thompson, *War and Government in Habsburg Spain, 1560–1620* (1976).
4. J. Inglis-Jones, 'The Battle of the Dunes, 1658: Condé, War and Power Politics', *War in History*, 1 (1994), pp. 260–1.
5. M. Fissel, *English Warfare, 1511–1642* (2001), pp. 181–206.
6. M. Fissel, *The Bishops' Wars: Charles I's Campaigns Against Scotland, 1638–1640* (Cambridge, 1994), pp. 62–151.
7. John Chetwynd, envoy in Turin, to Charles Hedges, 13 Nov. 1706, BL. Add. 61525 fol. 8.
8. F. Ansani, '"This French artillery is very good and very effective." Hypotheses on the Diffusion of a new Military Technology in Renaissance Italy," *JMH*, 83 (2019), pp. 359–76.
9. F. Ansani, 'Supplying the army, 1498. The Florentine campaign in the Pisan countryside', *Journal of Medieval Military History*, 17 (2019).
10. M. Elvin, *The Pattern of the Chinese Past* (1973), pp. 95–7.
11. A. Waldron, *The Great Wall of China: From History to Myth* (Cambridge, 1990), pp. 125–39.
12. F. Mote, 'The Tu-Mu Incident of 1449', in F. Kierman and J.K. Fairbank (eds), *Chinese Ways in Warfare* (Cambridge, Mass., 1974), pp. 243–72.
13. E.L. Dreyer, 'Zhao Chongguo: A Professional Soldier of the Former Han Dynasty', *JMH*, 72 (2008), pp. 665–725, esp. 710.
14. G. Ágoston, *Guns for the Sultan: Military Power and the Weapons Industry in the Ottoman Empire* (Cambridge, 2005); G. Veinstein, 'Some Views on Provisioning in the Hungarian Campaigns of Suleyman the Magnificent', in Veinstein (ed.), *Etat et Société dans l'Empire Ottoman, XVIe-XVIIIe siècles* (Aldershot, 1994), pp. 177–85.
15. C. Finkel, *The Administration of Warfare: Ottoman Campaigns in Hungary, 1593–1606* (Vienna, 1988).
16. G. Ágoston, 'Muslim-Christian Acculturation: Ottomans and Hungarians from the Fifteenth to the Seventeenth Centuries', in B. Bennassar and R. Sauzet (eds), *Chrétiens et Musulmans à la Renaissance* (Paris, 1994), p. 296.

17. V. Aksan, 'Ottoman war and warfare 1453–1812', in J. Black (ed.), *War in the early modern world* (1999), pp. 156 & 160–1.
18. R. Law, *The Horse in West African History* (Oxford, 1980).
19. J. Thornton, 'Warfare, slave trading and European influence: Atlantic Africa 1450–1800', in J. Black (ed.), *War in the early modern world* (1999), pp. 140–2.
20. P. Crone, *Slaves on Horses: The Evolution of the Islamic Polity* (Cambridge, 1980).
21. H. Kamen, *The Duke of Alba* (New Haven, Conn., 2004), p. 19; G. Phillips, 'Strategy and Its Limitations: The Anglo-Scots Wars, 1480–1550', *War in History*, 6 (1999), p. 406.
22. A.N. Kurat, 'The Turkish Expedition to Astrakhan in 1569 and the Problem of the Don-Volga Canal', *Slavonic and East European Review*, 40 (1961), pp. 7–23.
23. W.E.D. Allen, *Problems of Turkish Power in the Sixteenth Century* (1963), pp. 36–7.
24. R. Murphey, 'A Comparative Look at Ottoman and Habsburg Resources and Readiness for War *c* 1520 to *c*. 1570', in G. Hernán and D. Maffi (eds), *Guerra y Sociedad en la Monarquía Hispánica* (2 vols, Madrid, 2006), I, pp. 76 & 102.
25. S. Morillo, 'Contrary Winds: Theories of History and the Limits of *Sachkritik*', in G.I. Halfond (ed.), *The Medieval Way of War* (Farnham, 2015), pp. 205–22.
26. R. Murphey, 'The Garrison and its Hinterland in the Ottoman East, 1578–1605', in A.C.S. Peacock (ed.), *The Frontiers of the Ottoman World* (Oxford, 2009), p. 369.
27. Allen, *Problems*, p. 38.
28. A. Burton, *The Bukharans: A Dynastic, Diplomatic and Commercial History, 1550–1702* (New York, 1997), p. 117.
29. S. Pepper, 'The siege of Siena in its International Context', in M. Ascheri, G. Mazzoni, and F. Nevola (eds), *L'Ultimo Secolo della Republica di Siena* (Siena, 2008), p. 466.
30. M. Greene, 'The Ottomans in the Mediterranean', in V. Aksan and D. Goffman (eds), *The Early Modern Ottomans: Remapping the Empire* (Cambridge, 2007), p. 110.
31. J. Black, 'The Limits of Empire: The Case of Britain', in T. Andrada and W. Reger (eds), *The Limits of Empire: European Imperial Formations in Early Modern World History* (Farnham, 2012), pp. 175–81.
32. K.M. Swope, 'Civil-Military Coordination in the Bozhou Campaign of the Wanli Era', *War and Society*, 18, 2 (2000), pp. 49–70.
33. M. Abir, *Ethiopia and the Red Sea* (1980), p. 136.
34. D. Grummitt, 'Flodden 1513: Re-examining British Warfare at the End of the Middle Ages', *JMH*, 82 (2018), p. 16.
35. G.E. Rothenburg, *The Austrian Military Border in Croatia, 1522–1747* (Urbana, Ill., 1960); G. Pálffy, *The Kingdom of Hungary and the Habsburg Monarchy in the Sixteenth Century* (Boulder, Col., 2009), pp. 114–17; J.D. Tracy, *Balkan Wars; Habsburg Croatia, Ottoman Bosnia, and Venetian Dalmatia, 1499–1617* (Lanham, MD., 2016).
36. G.E. Rothenberg, 'Christian Insurrections in Ottoman Dalmatia 1580–96', *Slavonic and East European Review*, 40 (1961), p. 141.
37. C. Finkel, 'The Costs of Ottoman Warfare and Defence', *Byzantinische Forschungen*, 16 (1990), p. 96.

38. P. Brummett, 'The River Crossing: Breaking Points (Metaphorical and "Real") in Ottoman Mutiny', in J. Hathaway (ed.), *Rebellion, Repression, Reinvention. Mutiny in Comparative Perspective* (Westport, Conn., 2001), pp. 222–3.
39. For Muscovy/Russia, B.L. Davies (ed.), *Warfare in Eastern Europe, 1500–1800* (Leiden, 2012).
40. For consent working, N. Younger, *War and Politics in the Elizabethan Counties* (Manchester, 2012).
41. P. Lenihan, *Confederate Catholics at War, 1641–49* (Cork, 2001).
42. S. Porter, *Destruction in the English Civil Wars* (1994).
43. P. Wilson, 'German Women and War, 1500–1800', *War in History*, 3 (1996), pp. 127–60, and 'The Military and Rural Society in the Early Modern Period', *German History*, 18 (2000), pp. 217–23; J. Lynn, *Women, Armies, and Warfare in early modern Europe* (Cambridge, 2008).
44. J. Glete, *Warfare at Sea, 1500–1650* (2000), pp. 186–7; J. Bruijn, 'States and Their Navies from the Late Sixteenth to the End of the Eighteenth Centuries', in P. Contamine (ed.), *War and Competition between States* (Oxford, 2000), pp. 78–9.
45. C.R. Phillips, *Six Galleons for the King of Spain: Imperial Defense in the Early Seventeenth Century* (Baltimore, MD, 1992).
46. A. Fuess, 'Rotting Ships and Razed Harbors: The Naval Policy of the Mamluks', *Mamlūk Studies Review*, 5 (2001), pp. 45–71, esp. p. 60.
47. P. Perez-Mallaina, *Spain's Men of the Sea: Daily Life on the Indies Fleets in the Sixteenth Century* (Baltimore, MD, 1998).
48. F. Bethencourt and D. Ramada Curto (eds), *Portuguese Oceanic Expansion, 1400–1800* (Cambridge, 2007).
49. J.F. Guilmartin, 'The Logistics of Warfare at Sea in the Sixteenth Century: The Spanish Perspective', in Lynn (ed.), *Feeding Mars*, pp. 109–36.
50. S. Özbaran, 'Ottoman naval policy in the south', in M. Kunt and C. Woodhead (eds), *Süleyman the Magnificent and his age: the Ottoman Empire in the early modern world* (1995), p. 64.
51. R. Murphey, *Ottoman Warfare, 1500–1700* (2000).
52. J. Gommans, *Mughal Warfare: Indian Frontiers and High Roads to Empire, 1500–1700* (2002).
53. J. Tracy, *Emperor Charles V, Impressario of War: Campaign Strategy, International Finance, and Domestic Politics* (Cambridge, 2002).
54. C. Finkel, 'The Cost of Ottoman Warfare and Defence', *Byzantinische Forschungen*, 16 (1990), pp. 91–103.
55. J.B. Wood, *The King's Army: Warfare, Soldiers and Society during the Wars of Religion in France, 1562–1576* (Cambridge, 1996); R. Knecht, *The French Civil Wars, 1562–1598* (Harlow, 2000).
56. C. Kapser, *Die bayerische Kriegorganisation in der zweiten Hälfte des dreissigjährigne Krieges 1635–1648/49* (Münster, 1997); P.H. Wilson, *Europe's Tragedy: A New History of the Thirty Years War* (2010), p. 623.
57. L. White, 'Strategic geography and the Spanish Habsburg monarchy's failure to recover Portugal, 1640–1668', *JMH*, 71 (2007), pp. 373–409.
58. P. Nath, *Climate of Conquest: Warfare, Environment, and Empire in Mughal North India* (New Delhi, 2019).

59. T. Ertman, *Birth of the Leviathan: Building States and Regimes in Medieval and Early Modern Europe* (Cambridge, 1997).

60. A.J. Nusbacher, 'Civil Supply in the Civil War: Supply of Victuals to the New Model Army on the Naseby Campaign, 1–14 June 1645', *English Historical Review*, 115 (2000), pp. 145–60, esp. pp. 159–60.

61. C. Tilly, *Coercion, Capital, and European States, AD 990–1992* (Oxford, 1992); S. Morillo, 'The Sword of Justice: War and State Formation in Comparative Perspective', *Journal of Medieval Military History*, 4 (2006), pp. 1–17.

62. L.E. Staiano-Daniels, 'Determining Early Modern Army Strength: The Case of Electoral Saxony', *JMH*, 83 (2019), pp. 1014–16.

63. J. Fynn-Paul, *War, Entrepreneurs and the State in Europe and the Mediterranean, 1300–1800* (Leiden, 2014).

64. G. Robinson, *Horses, People and Parliament in the English Civil War: Extracting Resources and Constructing Allegiance* (Farnham, 2012).

65. D. Parrott, 'France's War against the Habsburgs, 1624–1659: the Politics of Military Failure', in E. García Hernan and D. Maffi (eds), *Guerra y Sociedad en la Monarquía Hispánica: Política, Estrategia y Cultura en La Europa Moderna, 1500–1700* (2 vols, Madrid, 2006), I, p. 33.

66. F. Redlich, *The German Military Enterpriser and his Work Force, 14th to 18th Centuries* (Wiesbaden, 1964–5); I.A.A. Thompson, *War and Government in Habsburg Spain, 1560–1620* (1976); D. Goodman, *Spanish Naval Power, 1589–1665. Reconstruction and Defeat* (Cambridge, 1997), pp. 29–36.

67. D. Croxton, 'A territorial imperative? The military revolution, strategy and peacemaking in the Thirty Years' War', *War in History*, 5 (1998), pp. 253–79.

68. D. Parrott, *Richelieu's Army. War, Government and Society in France, 1624–42* (Cambridge, 2001). This is closer to the reality of weaknesses and failures than B. Kroener, *Les Routes et les Étapes: Die Versorgung der französischen Armeen in Nordostfrankreich, 1635–1661: Ein Beitrag zur Verwaltungsgeschichte des Ancien Régime* (Münster, 1980). For serious financial problems, R.J. Bonney, *The King's Debts, Finance and Politics in France, 1589–1661* (Oxford, 1981).

69. J.A. Lynn, 'How War Fed War: The Tax of Violence and Contributions during the Grand Siècle', *Journal of Modern History*, 65 (1993), pp. 286–310.

70. F. McArdle, *Altopasico: A Study in Tuscan Rural Society, 1587–1784* (1978).

71. R. Murphey, 'Ottoman Expansion, 1451–1566', in Mortimer (ed.), *Early Modern Military History*, p. 58.

Chapter 3

1. J.A. Lynn, 'The Evolution of Army Style in the Modern West, 800–2000', *International History Review*, 18 (1996), pp. 505–45.

2. J.A. Lynn, 'Food, Funds, and Fortresses: Resources Mobilization and Positional Warfare in the Campaigns of Louis XIV', in Lynn (ed.), *Feeding Mars*, pp. 137–60, *Giant of the Grand Siècle: The French Army, 1619–1715* (Cambridge, 1997) and *The Wars of Louis XIV* (Harlow, 1999), D. Dee, 'The Survival of France: Logistics and Strategy in the 1709 Flanders Campaign', JMH, 84 (2020), pp. 1021–50.

3. G. Perjés, 'Army Provisioning, Logistics and Strategy in the Second Half of the Seventeenth Century', *Acta Historica Academiae Scientarium Hungaricae*, 16,

no. 1/2 (1970), pp. 4–13; J. Milot, 'Un problème operationnel du XVIIe siècle illustré par un cas regional', *Revue du Nord*, 53 (1971), pp. 269–85.

4. T. Hale, *An Account of several New Inventions and Improvements now necessary for England* (1691).

5. C. Storrs, 'The Army of Lombardy and the Resilience of Spanish Power in Italy in the Reign of Carlos II, 1665–1700, Part II', *War in History*, 5 (1998), p. 4.

6. O. van Nimwegen, *De subsistentie van het leger: Logistiek en Strategie van het Geallierde en met name het staatse leger tijdens de spaanse successieoorlog in de Nederlanden en het Heilige Roomse Rijk, 1701–1712* (Amsterdam, 1995), with English summary.

7. W.D. Godsey, *The Sinews of Habsburg Power: Lower Austria as a Fiscal-Military State, 1650–1820* (Oxford, 2018).

8. C. Storrs and H.M. Scott, 'The Military Revolution and the European Nobility, *c.* 1600–1800', *War in History*, 3 (1996), p. 39; G. Rowlands, 'Louis XIV, Aristocratic Power and the Elite Units of the French Army', *French History*, 13 (1999), pp. 304 & 329–31.

9. M.C.'t Hart, *The Making of a Bourgeois State: War, Politics and Finance During the Dutch Revolt* (Manchester, 1993) and *The Dutch Wars of Independence. Warfare and Commerce in the Netherlands 1570–1680* (2014).

10. I. Gentles, *The New Model Army in England, Ireland and Scotland, 1645–1653* (Oxford, 1992); G. Robinson, 'Horse Supply and the Development of the New Model Army, 1642–1646', *War in History*, 15 (2008), pp. 121–40 and *Horses, People and Parliament in the English Civil War: Extracting Resources and Constructing Allegiance* (Farnham, 2012).

11. J.S. Wheeler, *Cromwell in Ireland* (Dublin, 1999).

12. R.I. Frost, *After the Deluge: Poland-Lithuania and the Second Northern War* (Cambridge, 1993).

13. NA. SP. 80/15 fol. 122, 80/16 fol. 6.

14. G. Satterfield, *Princes, Posts and Partisans: The Army of Louis XIV and Partisan Warfare in the Netherlands, 1673–1678* (Leiden, 2003). For an eighteenth-century focus, J. Kunísch, *Der Kleine Krieg. Studien zum Heerwesen des Absolutismus* (Wiesbaden, 1973).

15. Report of 17 July 1676, NA. SP. 84/202.

16. G. Rowlands, *The Dynastic State and the Army under Louis XIV: Royal Service and Private Interest, 1661–1701* (Cambridge, 2002).

17. NA. SP. 78/142 fol. 289. For the most scholarly recent account, J.A. Lynn, 'Revisiting the Great Fact of War and Bourbon Absolutism: The Growth of the French Army during the *Grand Siècle*', in Hernán and Maffi (eds), *Guerra y Sociedad*, I, pp. 49–61.

18. J. Nouzille, 'Charles V de Lorraine, les Habsbourg et la guerre contre les Turcs de 1683 à 1687', in J.P. Bled, E. Faucher and R. Tavenaux (eds), *Les Habsbourg et la Lorraine* (Nancy, 1988), p. 112.

19. T.M. Barker, 'New Perspectives on the Historical Significance of the "Year of the Turk"', *Austrian History Yearbook*, 19–20 (1983–4), p. 8.

20. C. Karges, 'The Logistics of the Allied War Effort in the Mediterranean', *Mitteilungen des* Österreichischen *Staatsarchiv Sonderband*, 16 (2018), pp. 95–

118; A. Graham and P. Walsh (eds), *The British Fiscal-Military States 1660–c. 1783* (2016).

21. J.A. Millward, *Beyond the Pass: Economy, Ethnicity, and Empire in Qing Central Asia, 1759–1864* (Stanford, Calif., 1998).

22. T. Saguchi, 'The Formation of the Turfan principality under the Qing Empire', *Acta Asiatica*, 41 (1981), pp. 76–94.

23. R. Hellie, 'The Petrine Army: Continuity, Change, and Impact', *Canadian-American Slavic Studies*, 8 (1974), p. 250.

24. Cambridge, University Library, Add. 6570.

25. A.D. Francis, *The First Peninsular War, 1702–13* (1975), pp. 339–42; Karges, 'Logistics', pp. 113–14.

26. G. Rowlands, *The Financial Decline of a Great Power: War, Influence and Money in Louis XIV's France* (Oxford, 2012).

27. H. Drévillon, *L'impôt du sang: Les officiers dans l'armée de Louis XIV, 1661–1715* (Paris, 2006); R. Blaufarb, 'Noble Privilege and Absolutist State Building: French Military Administration after the Seven Years' War', *French Historical Studies*, 24 (2001), pp. 238–9; W. Beik, 'The Absolutism of Louis XIV as Social Collaboration', *Past and Present*, no. 188 Aug. 2005, pp. 195–224; army pp. 212–18; O. Chaline, *Les armées du Roi: Le grand chantier XVIIe-XVIIIe siècle* (Paris, 2016).

28. G. Rowlands, 'Moving Mars: The Logistical Geography of Louis XIV's France', *French History*, 25 (2011), pp. 492–514.

Chapter 4

1. Yorke to Edward Weston, 23 Aug. 1763, BL. Add. 58213 fols 282–3.

2. P.C. Perdue, 'Military Mobilization in Seventeenth- and Eighteenth-century China, Russia, and Mongolia', *Modern Asian Studies*, 30 (1996), pp. 757–93, esp. 777–81.

3. P.C. Perdue, *China Marches West: The Qing Conquests of Central Eurasia* (Cambridge, Mass., 2005), pp. 270–89.

4. Perdue to Black, undated letter.

5. Y. Dai, 'The Qing State, Merchants, and the Military Labor Force in the Jinchuan Campaigns', *Late Imperial China*, 22 (2001), pp. 35–90.

6. Y. Dai, '*Yingyun Shengxi*: Military Entrepreneurship in the High Qing Period: 1700–1800', *Late Imperial China*, 26, no. 2 (2005), pp. 1–67 and 'Military Finance in the High Qing Period: An Overview', in N. DiCosmo (ed.), *Military Culture in Imperial China* (Cambridge, Mass., 2009), pp. 296–316, 380–2.

7. G.H. Luce, 'Chinese Invasions of Burma in the Eighteenth Century', *Journal of the Burma Research Society*, 15 (1925), pp. 115–28; J.K. Jung, 'The Sino-Burmese War, 1766–1770: War and Peace under the Tributary System', *China Papers*, 24 (1971), pp. 74–103; Y. Dai, 'A Disguised Defeat. The Myanmar Campaign of the Qing Dynasty', *Modern Asian Studies*, 38 (2004), pp. 145–89.

8. A. Graham, *Corruption, Party and Government in Britain, 1702–1713* (Oxford, 2015).

9. P.G.M. Dickson, *The Financial Revolution in England: A Study in the Development of Public Credit 1688–1756* (1967); C. Wilkinson, *The British Navy*

and the State in the Eighteenth Century (Woodbridge, 2004); A. Graham, 'The British Financial Revolution and the empire of credit in St Kitts and Nevis, 1706–21', *Historical Research*, 91 (2018), pp. 685–704.

10. R. Morriss, *Naval Power and British Culture, 1760–1850: Public Trust and Government Ideology* (Farnham, 2004) and *The Foundations of British Maritime Ascendancy: Resources, Logistics and the State, 1755–1815* (Cambridge, 2011).

11. A.R. Wadia, *The Bombay Dockyard and the Wadia Master Builders* (Bombay, 1957).

12. D. Syrett, *Shipping and Military Power in the Seven Years War: The Sails of Victory* (Exeter, 2008); R.J.W. Knight and M. Wilcox, *Sustaining the Fleet, 1793–1815: War, the British Navy and the Contractor State* (Woodbridge, 2010); Wilcox, '"This Great Complex Concern": Victualling the Royal Navy on the East Indies Station, 1780–1815', *Mariner's Mirror*, 97 (2011), pp. 32–49. For the Spanish counterpart, R. T. Sánches, *Military Entrepreneurs and the Spanish Contractor State in the Eighteenth Century* (Oxford, 2016).

13. R. Middleton (ed.), *Amherst and the Conquest of Canada* (Stroud, 2003); D.R. Cubbison, *All Canada in the Hands of the British: General Jeffery Amherst and the 1760 Campaign to Conquer New France* (Norman, Ok., 2014).

14. L. Lockhart, *Nadir Shah* (1938); J. Sarkar, *Nadir Shah in India* (Calcutta, 1973); M. Axworthy, *Sword of Persia: Nader Shah, from Tribal Warrior to Conquering Tyrant* (2006) and 'The Army of Nader Shah', *Iranian Studies*, 40 (2007), pp. 635–46.

15. A.E. Tucker, *Nadir Shah's Quest for Legitimacy in Post-Safavid Iran* (Gainesville, Fl., 2006).

16. The Ottomans are referred to as Turks from here on.

17. V. Aksan, 'Ottoman Military Power in the Eighteenth Century', in B. Davies (ed.), *Warfare in Eastern Europe 1500–1800* (Leiden, 2012), p. 321.

18. G.J. Bryant, 'British Logistics and the Conduct of the Carnatic Wars, 1746–1783', *War in History*, 11 (2004), pp. 278–306, and 'Asymmetric Warfare: The British Experience in Eighteenth-Century India', *JMH*, 68 (2004), pp. 431–69.

19. Cornwallis to Medows, 4 Jan. 1791, NA. PRO. 30/11/173 fols 43, 45.

20. Cornwallis to Medows, 28 Dec. 1790, NA. PRO. 30/11/173 fol. 38.

21. R.G.S. Cooper, *The Anglo-Maratha Campaigns and the Contest for India: The Struggle for Control of the South Asian Military Economy* (Cambridge, 2003), p. 65.

22. V.H. Aksan, 'Feeding the Ottoman troops on the Danube, 1768–1774', *War and Society*, 13 (1995), pp. 1–14; G. Ágoston, 'Ottoman Warfare in Europe 1453–1826', in J. Black (ed.), *European Warfare 1453–1815* (Basingstoke, 1999), p. 142.

23. G. Ágoston, *Guns for the Sultan: Military Power and the Weapons Industry in the Ottoman Empire* (Cambridge, 2005), p. 199; B.W. Menning, 'Russian Military Innovation in the Second Half of the Eighteenth Century', *War and Society*, 2 (1984), pp. 33–5.

24. Z. Kohut, *Russian Centralism and Ukrainian Autonomy* (Cambridge, Mass., 1988), pp. 104–24; B.W. Menning, 'G.A. Potemkin and A.I. Chernyshev: Two Dimensions of Reform and the Military Frontier in Imperial Russia', *Consortium on Revolutionary Europe. Proceedings*, I (1980), pp. 241–3.

25. V.H. Aksan, 'The One-Eyed Fighting the Blind: Mobilisation, Supply, and Command in the Russo-Turkish War of 1768–1774', *International History*

Review, 15 (1993), pp. 221–38 and *Ottoman Wars, 1700–1870: An Empire Besieged* (2007); B.L. Davies, *The Russo-Turkish War, 1768–1774: Catherine II and the Ottoman Empire* (2016).

26. J. Keep, 'Feeding the Troops: Russian Army Supply Policies During the Seven Years' War', *Canadian Slavonic Papers*, 29 (1987), pp. 24–44, quotes 24, 43.

27. J. Luh, '"Strategie und Taktck" in Ancien Régime', *Militargeschichtliche Zeitschrift*, 64 (2005), pp. 101–31.

28. G. Bannerman, *Merchants and the Military in Eighteenth-Century Britain: British Army Contracts and Domestic Supply, 1739–1763* (2008).

29. Royal Archives, Cumberland Papers, 7/279, 288, 291, 306; Durham, CRO. D/Lo/F/745/57; C.J. Terry (ed.), *The Albemarle Papers* (2 vols, Aberdeen, 1902), I, p. 114.

30. Chatsworth, papers of 3rd Duke, History of Parliament transcripts.

31. See also Michael Hatton to Robert, 4th Earl of Holdernesse, 18 July, 13 Aug. 1758, BL. Eg. 3443 fols 34, 42, 48.

32. Bod. MS. Eng. Hist. c.314 fols 46, 51. Current value about £8.5 million.

33. Sir John Ligonier to Cumberland, 24 July 1746, RA. CP. 17/252; Edmund Martin to 2nd Duke of Richmond, 19 Aug. 1748, Goodwood Mss. 107 no. 685; Captain William Fawcett to James Lister, 24 Oct. 1760, Halifax, Calderdale District Archives, SH: 7/FAW/60; Major-General George Townshend to 3rd Earl of Bute, 17 Sept. 1761, Mount Stuart, Bute papers 7/23.

34. E. Lund, *War for the Every Day: Generals, Knowledge and Warfare in Early Modern Europe, 1680–1740* (Westport, Conn., 1999).

35. For the French army in this period, Françolis de Chenevières, *Détails militires dont la connoissance est nécessaires à tous les officiers, et principalement aux commissaires des Guerres* (Paris, 1750); C-N. Dublanchy, *Une intendance d'armée au XVIII siècle. Etudes sur les services administratifs à l'armée de Soubise pendant la Guerre de Sept Ans, d'après la correspondence et les papiers inédit de l'intendant François-Marie Gayot* (Paris, 1908).

36. Vienna, Haus-, Hof-, und Staatsarchiv, Staatskanzlei, England, Noten 2.

37. Brigadier-General James Cholmondely to George, 3rd Earl of Cholmondely, 19 Nov. 1745, Chester, CRO. DCH/X/9a/11.

38. Wade to Henry, 3rd Viscount Lonsdale, 13 Dec. 1745, Carlisle, Cumbria CRO. D/Pen Acc 2689.

39. Thomas, Duke of Newcastle, Secretary of State for the Southern Department, to Charles, 2nd Viscount Townshend, his counterpart for the Northern Department, 27 June 1729, NA. SP. 43/78.

40. De Lancey to Sir Thomas Robinson, Secretary of State for the Southern Department, 7 Aug. 1755, BL. Add. 32858 fol. 25.

41. Wolfe to Colonel Charles Hotham, 9 Aug. 1758, Hull, University Library, Hotham papers, DDHo/4/7.

42. M.P. Gutmann, *War and Rural Life in the Early Modern Low Countries* (Princeton, NJ., 1980); C.R. Friedrichs, *Urban Society in an Age of War: Nördlingen, 1580–1720* (Princeton, NJ, 1979).

43. HL., Loudoun papers, no. 10125; James, Lord Tyrawly, British commander, to Marquis of Pombal, Portuguese First Minister, 24 July 1762, Belfast, Public Record Office of Northern Ireland, T 2812/8/48.

44. R.K. Showman (ed.), *The Papers of General Nathanael Greene*, II (Chapel Hill, N.C., 1980), p. 135.
45. Greene to North Carolina Board of War, 14 Dec. 1780, Washington, Library of Congress, Greene Letterbook; E.W. Carp, *To Starve the Army at Pleasure: Continental Army Administration and American Political Culture, 1775–1783* (Chapel Hill, NC., 1984); J. Hutson, *Logistics of Liberty. American Services of Supply in the Revolutionary War and after* (Newark, NJ, 1991).
46. Greene to Colonel Edward Carrington, 4 Dec. 1780, in R.K. Showman (ed.), *The Papers of General Nathanael Greene*, VI (Chapel Hill, NC., 1991), p. 517.
47. H.C. Syrett and J.E. Cooke (eds), *The Papers of Alexander Hamilton* (New York, 1961–79), II, p. 554.
48. W.E. Lee, 'Early American Ways of War. A new reconnaissance, 1600–1815', *Historical Journal*, 44 (2001), pp. 269–89.
49. R.M. Dunkerly and I.B. Boland, *Eutaw Springs: The Final Battle of the American Revolution's Southern Campaign* (Columbia, S.C., 2017).
50. R.A. Bowler, *Logistics and the Failure of the British Army in America, 1775–1783* (Princeton, NJ., 1975), pp. 231, 234.
51. W. Cobbett, *House of Commons*, XX, 794.
52. *The Papers of George Washington. Revolutionary War Series*, X, 230.
53. J. Shy, 'Logistical crisis and the American Revolution: a hypothesis', in Lynn (ed.), *Feeding Mars*, pp. 161–79.
54. D. Parrott, 'Cultures of Combat in the *Ancien Régime*. Linear warfare, noble values, and entrepreneurship', *International History Review*, 27 (2005), pp. 518–33.
55. C. Pichichero, *The Military Enlightenment: War and Culture in the French Empire from Louis XIV to Napoleon* (Ithaca, NY, 2017).

Chapter 5

1. J.P. LeDonne, 'Geopolitics, Logistics, and Grain. Russia's ambitions in the Black Sea Basin, 1737–1834', *International History Review*, 28 (2006), pp. 1–41.
2. Ainslie to Lord Grenville, Foreign Secretary, 26 Mar. 1793, NA. FO. 78/14 fols 38, 79–80.
3. G. Cole, *Arming the Royal Navy, 1793–1815: The Office of Ordnance and the State* (2012).
4. J. Macdonald, *Feeding Nelson's Navy: The true story of food at sea in the Georgian era* (2004) and *The British Navy's Victualling Board 1793–1815* (Woodbridge, 2010); J. Davey, *The Transformation of British Naval Strategy: Seapower and Supply in Northern Europe, 1808–1812* (Woodbridge, 2012).
5. Richmond to Major-General Charles Grey, 27 Ap. 1782, Durham, University Department of Palaeography, papers of 1st Earl Grey, no. 61.
6. D. French, *The British way in warfare 1688–2000* (1990), pp. 91, 117.
7. N.A.M. Rodger, *The Command of the Ocean: A Naval History of Britain II: 1649–1815* (2004), p. 639.
8. Knight, *Britain*, pp. 437–8.
9. *Hansard's Parliamentary Debates*, 26 (1813), columns 23–4.

10. J.M. Hartley, *Russia, 1762–1825: Military Power, the State and the People* (Westport, Conn., 2008).

11. Castlereagh to Colonel J.W. Gordon, 17 Oct. 1805, 20 Nov. 1808, BL. Add. 49480 fols 6, 58–9.

12. Knight, *Britain*, pp. 410, 422–3 & 425.

13. NA. WO. 6/35, pp. 118–19, 5, 17, 331, 54–9 & 75–9; C.D. Hall, *British Strategy in the Napoleonic War 1803–1815* (Manchester, 1992), pp. 20–1; F.O. Cetre, 'Beresford and the Portuguese army, 1809–1814', in A.D. Berkeley (ed.), *New Lights on the Peninsular War* (Almada, 1991), pp. 149–56.

14. G.J. Bryant, 'Asymmetric Warfare: The British Experience in Eighteenth-Century India', *JMH*, 68 (2004), p. 468.

15. W. Reid, 'Tracing the Biscuit: The Commissariat in the Peninsular War', *Militaria. Revista de Cultura Militar*, 7 (1995), pp. 101–8.

16. Knight, *Britain*, pp. 427–9.

17. Ibid., pp. 109–70.

18. R. Sutcliffe, *British Expeditionary Warfare and the Defeat of Napoleon, 1793–1815* (Woodbridge, 2016).

19. M. Duffy, *Soldiers, Sugar and Seapower: The British Expeditions to the West Indies and the War Against Revolutionary France* (Oxford, 1987), pp. 190–1.

20. Colonel John Moore to his father, John, 16 Mar. 1801, BL. Add. 59281 fol. 69.

21. J. Gurwood (ed.), *The Dispatches of Field Marshal, the Duke of Wellington* (12 vols, 1837–8), pp. IX–X, 162, 363 & 479–80.

22. R.W. Hamilton (eds.), *Letters and Papers of Sir Thomas Byam Martin* (1898), pp. II & 409.

23. Memorandum of 17 Dec. 1808, Exeter, CRO., Guard's letterbook, 49/33 fol. 10; I. Robertson, *A Commanding Presence: Wellington in the Peninsula, 1808–1814 – Logistics, Strategy, Survival* (Stroud, 2008).

24. 'An Ohio Volunteer', *The Capitulation* (Chillicothe, 1812), in *War on the Detroit*, edited by M.M. Quaife (Chicago, Ill., 1940), p. 209.

25. J.C.A. Stagg (ed.), *The Papers of James Madison, The Presidential Series* (Charlottesville, Va., 2004) vol. 279.

26. *Madison*, vol. 627.

27. P. Wetzler, *War and Subsistence: The Sambre and Meuse Army in 1794* (New York, 1985); A. Forrest, 'The Logistics of Revolutionary War in France', in R. Chickering and S. Förster (eds), *War in an Age of Revolution, 1775–1815* (Cambridge, 2010), pp. 177–96; J.R. Hayworth, *Revolutionary France's War of Conquest in the Rhineland* (Cambridge, 2019).

28. T.C.W. Blanning, *The French Revolution in Germany: Occupation and Resistance in the Rhineland, 1792–1802* (Oxford, 1983) and *The French Revolutionary Wars 1787–1802* (1996), pp. 158–69.

29. NAM., 1975–09–62-1.

30. C.J. Esdaile, 'De-constructing the French Wars. Napoleon as anti-strategist', *Journal of Strategic Studies*, 31 (2008), pp. 515–52.

31. Van Creveld, *Supplying War*, pp. 40–74.

32. J.T. Kuehn, *Napoleonic Warfare: The Operational Art of the Great Campaigns* (Santa Barbara, Calif., 2015).

33. K. Aalestad and J. Joor (eds), *Revisiting Napoleon's Continental System: Local, Regional, and European Experiences* (Basingstoke, 2015).

34. M. Broers, *The Napoleonic Empire in Italy, 1796–1814* (Basingstoke, 2005); A. Grab, 'Army, State and Society: Conscription and Desertion in Napoleonic Italy, 1802–1814', *Journal of Modern History*, 67 (1995), pp. 25–54.

35. A. Mikaberidze, *The Napoleonic Wars. A Global History* (Oxford, 2020), p. 531.

36. C. von Clausewitz, *On War*, edited by M. Howard and P. Paret (Princeton, NJ, 1976), p. 340.

37. M. Coreia de Andrade, 'The social and ethnic significance of the War of the Cabanos', in R.H. Chilcote (ed.), *Protest and Resistance in Angola and Brazil* (Berkeley, Calif., 1972), esp. pp. 98–103.

38. M. Santiago, *A Bad Peace and a Good War: Spain and the Mescalero Apache Uprising of 1795–1799* (Norman, Ok., 2018); H. Van de Ven, 'Military Mobilisation in China, 1840–1949', in J. Black (ed.), *War in the Modern World since 1815* (2003), pp. 25–6.

39. A. Peskin, *Winfield Scott and the Profession of Arms* (Kent, OH, 2004).

40. T.D. Johnson, *A Gallant Little Army: The Mexico City Campaign* (Lawrence, KS, 2007); P. Guardino, *The Dead March: A History of the Mexican-American War* (Cambridge, Mass., 2017).

41. D. Narrett, *Adventurism and Empire: The Struggle for Mastery in the Louisiana-Florida Borderlands, 1762–1803* (Chapel Hill, NC, 2015).

42. *Atlas of Portsmouth* (Portsmouth, 1975), section 3/6(d).

43. I.C. Hope, *A Scientific Way of War: Antebellum Military Science, West Point, and the Origins of American Military Thought* (Lincoln, NB, 2015).

44. J.R. Arnold and R.R. Reinertsen, *Crisis in the Snows: Russia Confronts Napoleon, The Eylau Campaign, 1806–1807* (Lexington, VA, 2007).

45. J. Morgan, 'War Feeding War? The Impact of Logistics on the Napoleonic Occupation of Catalonia', *JMH*, 73 (2009), pp. 83–116.

Chapter 6

1. A. Lambert, *The Crimean War. British Grand Strategy against Russia, 1853–56* (2nd edn, Farnham, 2011), pp. 349–50.

2. J. Sweetman, 'Military transport in the Crimean War, 1854–1856', *English Historical Review*, 88 (1973), pp. 81–98 and *War and Administration: The Significance of the Crimean War for the British Army* (Edinburgh, 1984).

3. F.C. Schneid, 'A Well-Coordinated Affair: Franco-Piedmontese War Planning in 1859', *JMH*, 76 (2012), pp. 395–425.

4. H. Mackinder, 'The Geographical Pivot of History', *Geographical Journal*, 23 (1904), pp. 421–37.

5. Graham to Raglan, 10 Jan. 1854, BL. Add. 79696 fol. 87.

6. NMM. Milne papers 124/2.

7. For a valuable cautious note, A.D. Harvey, 'Was the Civil War the First Modern War?', *History* (2012), pp. 272–80.

8. Lyons to Rear-Admiral Alexander Milne, Commander of the North American and West Indies station, 27 Feb. 1862, NMM. Milne papers 116/1.

9. R.G. Angevine, *The Railroad and the State: War, Politics, and Technology in Nineteenth-Century America* (Stanford, Calif., 2004).

10. R.C. Black, *The Railroads of the Confederacy* (Chapel Hill, NC., 1998).

11. T.B. Smith, *Grant Invades Tennessee: The 1862 Battles for Forts Henry and Donelson* (Lawrence, KS, 2016).

12. E. Hagerman, 'Field transportation and strategic mobility in the Union armies', *Civil War History*, 34 (1988), p. 171.

13. N.A. Trudeau, *Southern Storm: Sherman's March to the Sea* (New York, 2008).

14. E.J. Hess, *Civil War Logistics: A Study of Military Transportation* (Baton Rouge, LA., 2017).

15. M.M. Edling, *A Hercules in the Cradle: War, Money, and the American State, 1783–1867* (Chicago, Ill., 2014).

16. W.H. Roberts, *Civil War Ironclads: The U.S. Navy and Industrial Mobilization* (Baltimore, MD., 2002).

17. R.D. Goff, *Confederate Supply* (Durham, NC., 1969); W. Blair, *Virginia's Private War. Feeding Body and Soul in the Confederacy, 1861–1865* (New York, 2000).

18. D.G. Smith, '"Clear the Valley": The Shenandoah Valley and the Genesis of the Gettysburg Campaign', *JMH*, 74 (2010), pp. 1087–95.

19. K.M. Brown, *Retreat from Gettysburg: Lee, Logistics, and the Pennsylvania Campaign* (Chapel Hill, NC., 2005).

20. M.R. Wilson, *The Business of War: Military Mobilisation and the State, 1861–1865* (Baltimore, MD, 2006); D.W. Miller, *Second Only to Grant: Quartermaster General Montgomery C. Meigs* (Shippensburg, Penn., 2000); C.R. Newell and C.R. Shrader, 'The U.S. Army's Transition to Peace, 1865–66', *JMH*, 77 (2013), pp. 877–9.

21. B.H. Reid, *America's Civil War: The Operational Battlefield, 1861–1863* (2008).

22. D.E. Sutherland, 'Abraham Lincoln, John Pope, and the origins of total war', *JMH*, 56 (1992), pp. 581–2; B.F. Cooling, *Fort Donelson's Legacy: War and Society in Kentucky and Tennessee, 1862–1863* (Knoxville, Tn., 1997); J.W. Danielson, *War's Desolating Scourge: The Union's Occupation of North Alabama* (Lawrence, KS, 2012).

23. M. Grimsley, *The Hard Hand of War: Union Military Policy toward Southern Civilians, 1861–1865* (New York, 1995).

24. C. Royster, *The Destructive War: William Tecumseh Sherman, Stonewall Jackson, and the Americans* (New York, 1991).

25. E.A. Pratt, *The Rise of Rail-Power in War and Conquest, 1833–1914* (1915); D. Showalter, *Railroads and Rifles: Soldiers. Technology and the Unification of Germany* (Hamden, Conn., 1975); M.V. Creveld, *Supplying War*, pp. 82–108; A. Bucholz, *Moltke and the German Wars, 1864–1871* (Basingstoke, 2001).

26. I.F.W. Beckett (ed.), *Wolseley and Ashanti: the Ashanti War Journal and Correspondence of Major-General Sir Garnet Wolseley, 1873–1874* (2009).

27. J. Laband, 'The Slave Soldiers of Africa', *JMH*, 81 (2017), pp. 9–38.

28. S. Badsey, *Doctrine and Reform in the British Cavalry 1880–1918* (Aldershot, 2008).

29. B.A. Elleman and S. Kotkin, *Manchurian Railways and the Opening of China: An International History* (Armonk, NY, 2010).

30. F. Patrikeeff and H. Shukman, *Railways and the Russo-Japanese War: Transporting War* (Abingdon, 2009).

31. T.G. Otte and K. Neilson (eds), *Railways and International Politics: Paths of Empire, 1848–1945* (2006).

32. H. Van de Ven, 'Military Mobilisation in China, 1840–1949', in J. Black (ed.), *War in the Modern World since 1815* (2003), pp. 25–6.

33. R. Beal and R. Macleod, *Prairie Fire: The North-West Rebellion of 1885* (Edmonton, 1984).

34. I.J. Kerr, *Building the Railways of the Raj, 1850–1900* (Oxford, 1995).

35. D. Headrick, *When Information Came of Age: Technologies of Knowledge in the Age of Reason and Revolution, 1700–1850* (Oxford, 2000).

36. P.A. Shulman, '"Science can never demobilize": The United States navy and petroleum geology, 1898–1924', *History and Technology*, 19 (2003), pp. 367–71.

37. P.A. Shulman, *Coal and Empire: The Birth of Energy Security in Industrial America* (Baltimore, MD, 2015).

38. T.C. Winegard, *The First World Oil War* (Toronto, 2016).

39. R.A. Caulk, 'Armies as Predators: Soldiers and Peasants in Ethiopia c. 1850–1935', *International Journal of African Historical Studies*, 11 (1978), pp. 457–93; J. Lamphear, 'Sub-Saharan African Warfare', in J. Black (ed.), *War in the Modern World since 1815* (2003), pp. 171 & 177–8.

40. J. Dunn, '"For God, Emperor, and Country!" The Evolution of Ethiopia's Nineteenth-Century Army', *War in History*, 1 (1994), pp. 287–9.

41. Churchill, *The River War* (1899), p. vii.

42. D. Peers, 'South Asia', in J. Black (ed.), *War in the Modern World since 1815* (2003), p. 43.

43. J.R. Winkler, *Nexus* (Cambridge, Mass., 2008).

44. H. Kraay and T.L. Whigham (eds), *I Die With My Country: Perspectives on the Paraguayan War, 1864–1870* (Lincoln, NB, 2005).

45. A.W. Quiroz, 'Loyalist overkill: The socioeconomic costs of "repressing" the separatist insurrection in Cuba, 1868–1878', *Hispanic American Historical Review*, 78 (1998), p. 269.

46. W.F. Sater, *Andean Tragedy: Fighting the War of the Pacific, 1879–1884* (Lincoln, NB, 2007).

47. R.E. Paul, *Blue Water Creek and the First Sioux War, 1854–1856* (Norman, Ok,. 2004).

48. W.F. Sater, *Andean Tragedy. Fighting the War of the Pacific, 1879–1884* (Lincoln, NB., 2007).

49. J.W. Fortescue, *The Royal Army Service Corps: a history of transport and supply in the British Army* (1930).

50. D. Headrick, *The Tools of Empire: Technology and European Imperialism in the Nineteenth Century* (Oxford, 1981).

51. C.J. Nolan, *The Allure of Battle: A History of How Wars Have Been Won and Lost* (Oxford, 2017).

52. L. Bencze, *The Occupation of Bosnia and Herzegovina 1878* (Highland Lakes, NJ, 2005).

53. B. Collins, 'Defining victory in Victorian Warfare, 1860–1882', *JMH*, 77 (2013), pp. 905–6.

54. J.W. Steinberg (ed.), *The Russo-Japanese War in Global Perspective: World War Zero* (Leiden, 2005).

55. H. Bailes, 'Technology and imperialism: A case study of the Victorian army in Africa', *Victorian Studies*, 24 (1980), pp. 86, 102; I.F.W. Beckett, *The Victorians at War* (2003), pp. 197–8.

56. W.M. McBride, 'Strategic determinism in technology selection: The electric battleship and US naval-industrial relations', *Technology and Culture*, 33 (1992), p. 249.

Chapter 7

1. Germany, Austro-Hungary (Austria for short), Bulgaria and Turkey were allied, and the first three took part in these conquests.

2. L. Müller, 'The Swedish Convoy Office and Shipping Protection Costs', in A.M. Forssberg, M. Hallenberg, O. Husz and J. Nordin (eds), *Organising History. Studies in Honour of Jan Glete* (Lund, 2011), pp. 255–75.

3. J.F. Godfrey, *Capitalism at War: Industrial Policy and Bureaucracy in France, 1914–1918* (Oxford, 1987).

4. R. Dunley, *Britain and the Mine, 1900–1915: Culture, Strategy, and International Law* (Basingstoke, 2018); J. Goldrick, *After Jutland: The Naval War in Northern European Waters, June 1916–November 1918* (Annapolis, MD, 2018).

5. 18 [not 28 as in catalogue] Aug. 1915, LH. Fuller papers, IV/3/155.

6. *Daily Telegraph*, 29 Sept. 1915. The attack failed.

7. D. Aubin and C. Goldstein (eds), *The War of Guns and Mathematics: Mathematical Practices and Communities in France and Its Western Allies around World War I* (Providence, RI, 2014).

8. I have benefited from the advice of Anthony Saunders.

9. R. Duffett, *The Stomach For Fighting: Food and the Soldiers of the Great War* (Manchester, 2015).

10. Monash to Hannah, 30 May 1915, AWM, 3 DRL/2316, 1/1 p. 72.

11. J. Macdonald, *Supplying the British Army in the First World War* (Barnsley, 2019).

12. G.A. Tunstall, *Written in Blood: The Battles for Fortress Przemyśl in WWI* (Bloomington, Ind., 2016).

13. P. Dye, *The Bridge to Airpower: Logistics Support for the Royal Flying Corps Operations on the Western Front, 1914–1918* (Annapolis, MD., 2015).

14. J.P. Harris, *Men, Ideas and Tanks: British Military Thought and Armoured Forces, 1903–1939* (Manchester, 1995); B. Hammond, *Cambrai 1917: The Myth of the First Great Tank Battle* (2008).

15. R.B. Bruce, *A Fraternity of Arms: America and France in the Great War* (Lawrence, KS, 2003); A. Offer, *The First World War: Agrarian Interpretation* (Oxford, 1989).

16. I.M. Brown, *British Logistics on the Western Front, 1914–1919* (Westport, Conn., 1998).

17. D.R. Stone, *The Russian Army in the Great War: The Eastern Front, 1914–1917* (Lawrence, KS., 2015).

18. M.B. Barrett, *Prelude to Blitzkrieg: The 1916 Austro-German Campaign in Romania* (Bloomington, Ind., 2013).
19. E.J. Erickson, *Gallipoli: The Ottoman Campaign* (Barnsley, 2010); Y. Akin, *When the War Came Home: The Ottomans' Great War and the Devastation of an Empire* (Stanford, Calif., 2018).
20. R. Anderson, *The Forgotten Front: The East African Campaign, 1914–1918* (Stroud, 2004).
21. Sidney Rogerson, Exeter, DRO. 5277 M/F3/34.
22. R. Lee, *British Battle Planning in 1916 and the Battle of Fromelles: A Case Study of Evolving Skill* (Farnham, 2015).
23. W. Philpott, *War of Attrition: Fighting the First World War* (2014).
24. J. Boff, *Winning and Losing on the Western Front: The British Third Army and the Defeat of Germany in 1918* (Cambridge, 2012).

Chapter 8

1. T. Balkelis, *War, Revolution, and Nation-Making in Lithuania, 1914–1923* (Oxford, 2018).
2. Enclosure in General Staff, 'The Situation in Turkey, 15th March, 1920', NA. CAB. 24/101 fol. 317.
3. *The Times*, 20 Jan. 1925; D.S. Woolman, *Rebels in the Rif: Abd el Krim and the Rif Rebellion* (Stanford, Calif., 1968).
4. *The Times*, 26 Sept. 1924.
5. *The Times*, 6 Sept. 1924.
6. *The Times*, 3 Sept. 1927; H. van de Ven, *Warfare and Nationalism in China, 1925–1945* (2003).
7. General Staff, 'The situation in Turkey, 15 Mar. 1920', NA. CAB. 24/101/67 fol. 315.
8. A.P. Wavell, 'The Army and the Prophets', *RUSI Journal*, 75, no. 500 (1930), reprinted, vol. 155, no. 6 (Dec. 2010), p. 91.
9. J.T. Kuehn, *Agents of Innovation: The General Board and the Design of the Fleet That Defeated the Japanese Navy* (Annapolis, MD., 2008); N. Friedman, *Winning a Future War: War Gaming and Victory in the Pacific War* (Washington, 2017).
10. T.D. Biddle, *Rhetoric and Reality in Air Warfare: The Evolution of British and American Ideas about Strategic Bombing, 1914–1945* (Princeton, NJ., 2004).
11. K.E. Irish, 'Apt Pupil: Dwight Eisenhower and the 1930 Industrial Mobilisation Plan', *JMH*, 70 (2006), pp. 31–61.
12. M.A. Barnhart, *Japan Prepares for Total War: The Search for Economic Security, 1919–1941* (Ithaca, NY, 1987).
13. E. J. Drea, *Japan's Imperial Army: Its Rise and Fall, 1853–1945* (Lawrence, KS, 2009).
14. M. Peattie, E.J. Drea and H. van de Ven (eds),*The Battle for China: Essays on the Military History of the Sino-Japanese War of 1937–1945* (Stanford, Calif., 2011).
15. A.D. Coox, *Nomonhan: Japan Against Russia, 1939* (Stanford, Calif., 1985).
16. D.M. Glantz, *The Soviet Strategic Offensive in Manchuria, 1945: August Storm* (2003).

17. Secretary of State for War to Cabinet colleagues, 17 Jan. 1936, NA. CAB. 24/259.
18. M. Bernal, 'The Nghe-tinh Soviet Movement, 1930–1931', *Past and Present*, 92 (1981), pp. 148–68; T. Rettig, 'French Military Policies in the Aftermath of the Yên Bay Mutiny, 1930: Old Security Dilemmas Return to the Surface', *South East Asia Research*, 10 (2002), pp. 309–31; M. Thomas, 'Fighting "Communist banditry" in French Vietnam: The Rhetoric of Repression after the Yen Bay Uprising, 1930–1932', *French Historical Studies*, 33 (2010).
19. *The Times*, 28 Oct. 1933.
20. P.M. Ynsfran (ed.), *The Epic of the Chaco: Marshal Estigarribia's Memoir of the Chaco War, 1932–38* (Austin, TX., 1950); B. Farcau, *The Chaco War: Bolivia and Paraguay, 1932–1935* (Westport, Conn., 1996); M. Hughes, 'Logistics and the Chaco War: Bolivia versus Paraguay, 1932–1935', *JMH*, 69 (2005), pp. 411–37.
21. A. Guillamón, *Ready for Revolution: The CNT Defense Committees in Barcelona 1933–1938* (Oakland, Calif., 2014).
22. NA. WO. 106/1576, pp. 1–7.
23. NA. WO. 105/1580, pp. 2–7.
24. M. Hughes, 'The "European Aldershot" for the Second World War? The battle of the Ebro, 1938', *RUSI Journal*, 147, 6 (2002), pp. 76–81.
25. R.R. Reese, 'Lessons of the Winter War. A Study in the military effectiveness of the Red Army, 1939–1940', *JMH*, 72 (2008), pp. 825–52.

Chapter 9

1. R. Forczyk, *Case White. The Invasion of Poland, 1939* (Oxford, 2019), pp. 137–8, 213–14, 331.
2. G.P. Megargee, *Inside Hitler's High Command* (Lawrence, KS, 2000); G. Gross, *The Myth and Reality of German Warfare: Operational Thinking from Moltke the Elder to Heusinger* (Lexington, KY, 2016).
3. A.R. Jacobsen, *Miracle at the Litza: Hitler's First Defeat on the Eastern Front* (Philadelphia, Penn., 2017).
4. D. Stahel, *Operation Typhoon: Hitler's March on Moscow, October 1941* (Cambridge, 2013).
5. S.G. Fritz, *Ostkrieg: Hitler's War of Extermination in the East* (Lexington, KY, 2011).
6. P. Hayes, *Industry and Ideology: I.G. Farben in the Nazi Era* (Cambridge, 1987); M.T. Allen, *The Business of Genocide: The SS, Slave Labor, and the Concentration Camps* (Chapel Hill, NC, 2002); W. Gruner, *Jewish Forced Labor under the Nazis: Economic Needs and Racial Aims, 1938–1944* (Cambridge, 2006).
7. U. Herbert, *Hitler's Foreign Workers: Enforced Labour in Germany under the Third Reich* (Cambridge, 1977).
8. A. Stewart, *The First Victory: The Second World War and the East Africa Campaign* (New Haven, Conn., 2016).
9. K. Aizawa, 'Japanese Strategy in the First Phase of the Pacific War', in NIDS International Forum on War History, *Strategy in the Pacific War* (Tokyo, 2010), p. 37.

10. D.C. Fuquea, 'Advantage Japan: The Imperial Japanese Navy's Superior High Seas Refuelling Capability', *JMH*, 84 (2020), pp. 213–35.

11. D. Todman, *Britain's War. A New World, 1942–1947* (2020), p. 253.

12. M. Wilson, *Destructive Creation: American Business and the Winning of World War II* (Philadelphia, Penn., 2016).

13. C. Ogburn, *The Marauders* (New York, 1959), p. 141.

14. P. Kennedy, *Engineers of Victory: The Problem Solvers Who Turned the Tide in the Second World War* (New York, 2013).

15. P. Caddick-Adams, *Sand and Steel. A New History of D-Day* (2019), pp. 704–6.

16. R.C. Cooke and R.C. Nesbit, *Target: Hitler's Oil* (1985).

17. Blamey to Percy Spencer, Minister for the Army, 1 Sept. 1941, AWM. 3 DRL/6643, 1/2.

18. A. Harvey, 'The Bomber Offensive That Never Took Off: Italy's Regia Aeronautica in 1940', *RUSI Journal*, 154, no. 6 (Dec. 2009), pp. 99–100.

19. A. Toprani, 'The First Way for Oil: The Caucasus, German Strategy, and the Turning Point of the War on the Eastern Front, 1942', *JMH*, 80 (2016), pp. 815–54.

20. M. Geyer and A. Tooze (eds), *The Cambridge History of the Second World War. III. Total War: Economy, Society and Culture* (Cambridge, 2015).

21. P.S. Meilinger, 'A History of Effects-Based Air Operations', *JMH*, 71 (2007), pp. 144–59.

22. A.C. Mierzejewski, *The Collapse of the German War Economy, 1939–1945: Allied Air Power and the German National Railway* (Chapel Hill, NC, 1988).

23. C.L. Kolakowski, 'Gallantry, Courage, and Devotion to Duty. Merrill's Marauders in Burma', *Army History*, 116 (Summer 2020), pp. 6–23; R. Annett, *Drop Zone Burma: Adventures in Allied Air-Supply 1943–45* (Barnsley, 2008).

24. P. Nash, *The Development of Mobile Logistic Support in Anglo-American Naval Policy, 1900–1953* (Gainesville, Fl., 2009).

25. Ismay to Churchill, 16 Mar. 1944, LH, Alanbrooke 6/3/8.

26. C. Symonds, 'For Want of a Nail: The Impact of Shipping on Grand Strategy in World War II', *JMH*, 81 (2017), pp. 659–66.

27. H.H. Cox, *The General Who Wore Six Stars: The Inside Story of John C.H. Lee* (Lincoln, NB., 2018), S. C. Kepher, *COSSAC. Lt. Gen. Sir Frederick Morgan and the Genesis of Operation Overlord* (Annapolis, MD, 2020)

28. J. Prados, *Normandy Crucible* (New York, 2011).

29. J. Ludwig, *Rückzug: The German Retreat from France, 1944* (Lexington, KY, 2012).

30. P.P. O'Brien, *How the War Was Won: Air-Sea Power and Allied Victory in World War II* (Cambridge, 2015).

31. A.L. Gropman (ed.), *The Big 'L': American Logistics in World War II* (Washington, 1997).

32. C. Baxter, *The Secret History of RDX: The Super-Explosive that Helped Win World War II* (Lexington, KY, 2018).

33. A. Hill, *The Great Patriotic War of the Soviet Union, 1941–45: A Documentary Reader* (Abingdon, 2009), p. 172.

34. S.J. Hantzis, *Rails of War: Supplying the Americans and their Allies in China-Burma-India* (Lincoln, NB, 2017).

35. V. P. O'Hara, *Six Victories, North Africa, Malta, and the Mediterranean Convoy War, November 1941–March 1942* (Annapolis, MD, 2019).
36. U. Sundberg, *Re-supplying Tobruk 1941* (unpublished manuscript).
37. K. Roy, 'Military Loyalty in the Colonial Context: A Case Study of the Indian Army during World War II', *JMH*, 73 (2009), pp. 497–529.
38. *Ibid.*, p. 519.

Chapter 10

1. C.R. Shrader, *The First Helicopter War: Logistics and Mobility in Algeria, 1954–1962* (Westport, Conn., 1999) and *A War of Logistics. Parachutes and Porters in Indochina, 1945–1954* (Lexington, KY, 2015).
2. M.G. Clemis, *The Control War: The Struggle for South Vietnam, 1968–1975* (Norman, OK., 2018).
3. G.K. Van Nederveen, *USAF Airlift into the Heart of Darkness, the Congo, 1960–1978: Implications for Modern Air Mobility Planners* (Maxwell AFB, AL, 2001).
4. D. Tal, 'A Tested Alliance: The American Airlift to Israel in the 1973 Yom Kippur War', *Israel Studies*, 19, no. 3 (Fall 2014), pp. 29–54.
5. S. Abbott, 'Window on the World: Rebuilding Kabul International Airport in 2002', *RUSI Journal*, 150, no. 3 (June 2005), pp. 30–1.
6. R.C. Owen, *Air Mobility: A Brief History of the Air Experience* (Washington, 2013).
7. H.M. Tanner, *Where Chiang Kai-shek Lost China: The Liao-Shen Campaign* (Bloomington, Ind., 2015).
8. C.R. Shrader, *Communist Logistics in the Korean War* (Westport, Conn., 1995).
9. A. Bregman, *Israel's Wars. A history since 1947* (4th edn, 2016), pp. 24–9.
10. J.J. Tian Ser Seah, 'Singapore, Hong Kong, and the Royal Navy's War in Korea', *JMH*, 83 (2019), pp. 1213–34, and 84 (2020), pp. 237–60.
11. Dick Goodenough to Jeremy Black, email, 1 July 2020.
12. G.J. Veith, *Black April: The Fall of South Vietnam 1973–1975* (New York, 2012).
13. C.R. Shrader, *The Withered Vine: Logistics and the Communist Insurgency in Greece, 1945–1949* (Westport, Conn., 1999).
14. J.S. Brown, *Kevlar Legions: The Transformation of the U.S. Army, 1989–2005* (Washington, 2011).
15. Covering 1942–95, C.R. Shrader, *History of Operations Research in the United States Army* (3 vols, Washington, 2006–9).

Chapter 11

1. G.M. Wells, *Army River Crossing Doctrine and AirLand Battle Future: Applicable or Anachronistic?* (Fort Leavenworth, KS, 1990).
2. S. Said, *Legitimizing Military Rule: Indonesian Armed Forces Ideology, 1958–2000* (Jakarta, 2006); Z. Abul-Magd, *Militarizing the Nation: The Army, Business, and Revolution in Egypt* (Columbia, NY, 2017); P. Chambers and N. Waitoolkiat (eds), *Khaki Capital: The Political Economy of the Military in Southeast Asia: Economic Power 'out of the barrel of the gun.'* (Copenhagen, 2017).
3. J. Thomson, *Mercenaries, Pirates, and Sovereigns: State Building and Extraterritorial Violence in Early Modern Europe* (Princeton, NJ., 1994).

4. J. Holland, *Normandy '44. D-Day and the Battle for France* (2019), p. 278.
5. C.R. Shrader, *A War of Logistics: Parachutes and Porters in Indochina, 1945–1954* (Lexington, KY, 2015).
6. I. Trauschweizer, *The Cold War U.S. Army: Building Deterrence for Limited War* (Lawrence, KS, 2008).

Chapter 12
1. Memorandum 'Political situation of the countries to the east of the Bay of Bengal', in James Scott to Sir George Macartney, 28 Oct. 1785, BL. IO. G34/2.
2. W.J. Slim, *Defeat Into Victory* (London, 1956), pp. 28–30.
3. Cavan to Sir Frederick Maurice, 6 Feb. 1924, LH, Maurice papers, 3/5/150.
4. Maurice, *On the Uses of the Study of War* (1927), p. x.
5. A. Bucholz, *Moltke and the German Wars* (Basingstoke, 2001).
6. A.F. Wilt, *Food for War: Agriculture and Rearmament in Britain before the Second World War* (Oxford, 2001).
7. J. West, *Gunpowder, Government and War in the Mid-Eighteenth Century* (Woodbridge, 1991).
8. Montgomery, General Situation memorandum, 21 Jan. 1945, LH, Alanbrooke papers 6/2/37.
9. Rutgers University Library, New Brunswick, New Jersey, Fuller papers, Box 4, report on India, pp. 4, 48.
10. M. van Creveld, *Supplying War*, p. 235.
11. *Ibid.*, pp. 236–7.
12. *Ibid.*, p. 210.
13. M.D. Krause and R.D. Phillips (eds), *Historical Perspectives on the Operational Art* (Fort McNair, DC, 2005).

Selected Further Reading

Ágoston, G., *Guns for the Sultan: Military Power and the Weapons Industry in the Ottoman Empire* (Cambridge, 2005).

Aksan, V.H., *Ottoman Wars, 1700–1870: An Empire Besieged* (2007).

Bannerman, G., *Merchants and the Military in Eighteenth-Century Britain: British Army Contracts and Domestic Supply, 1739–1763* (2008).

Barrett, M.B., *Prelude to Blitzkrieg: The 1916 Austro-German Campaign in Romania* (Bloomington, Ind., 2013).

Black, J., *War in Europe: 1450 to the Present* (2016).

Bowler, R.A., *Logistics and the Failure of the British Army in America, 1775–1783* (Princeton, NJ., 1975).

Brewer, J., *The Sinews of Power: War, Money, and the English State, 1688–1783* (1989)

Brown, I.M., *British Logistics on the Western Front, 1914–1919* (Westport, Conn., 1998).

Carp, E.W., *To Starve the Army at Pleasure: Continental Army Administration and American Political Culture, 1775–1783* (Chapel Hill, NC., 1984).

Cole, G., *Arming the Royal Navy, 1793–1815: The Office of Ordnance and the State* (2012).

Cooper, R.G.S., *The Anglo-Maratha Campaigns and the Contest for India: The Struggle for Control of the South Asian Military Economy* (Cambridge, 2003).

Creveld, M. van, *Supplying War: Logistics from Wallenstein to Patton* (Cambridge, 1979).

Cubbison, D.R., *All Canada in the Hands of the British: General Jeffery Amherst and the 1760 Campaign to Conquer New France* (Norman, Ok., 2014).

Davey, J., *The Transformation of British Naval Strategy: Seapower and Supply in Northern Europe, 1808–1812* (Woodbridge, 2012).

Davies, B.L., *The Russo-Turkish War, 1768–1774: Catherine II and the Ottoman Empire* (2016).

Dennis, P. and J. Grey (eds), *Raise, Train and Sustain: Delivering Land Combat Power* (Canberra, 2010).

DiCosmo, N. (ed.), *Military Culture in Imperial China* (Cambridge, Mass., 2009).

Eccles, H., *Logistics in the National Defense* (Westport, Conn., 1981).

Fynn-Paul, J., (ed.), *War, Entrepreneurs and the State in Europe and the Mediterranean, 1300–1800* (Leiden, 2014).

Glete, J., *War and the State in Early Modern Europe: Spain, the Dutch Republic and Sweden as Fiscal-Military States, 1500–1600* (2002).

Huston, J.A., *The Sinew of War, Army Logistics, 1755–1953* (Washington, 1966).

Hutson, J., *Logistics of Liberty. American Services of Supply in the Revolutionary War and after* (Newark, NJ, 1991).

Kane, T., *Military Logistics and Strategic Performance* (2001).

Knight, R., and M. Wilcox, *Sustaining the Fleet, 1793–1815: War, the British Navy and the Contractor State* (Woodbridge, 2010).

Knight, R.J.W., *Britain against Napoleon: the Organisation of Victory, 1793–1815* (2013).

Lynn, J., *Feeding Mars: Logistics in Western Warfare from the Middle Ages to the Present* (Boulder, Col., 1993).

Macdonald, J., *Feeding Nelson's Navy: The true story of food at sea in the Georgian era* (2004).

Macdonald, J., *The British Navy's Victualling Board 1793–1815* (Woodbridge, 2010).

Maiolo, J., *Cry Havoc: How the Arms Race Drove the World to War 1931–1941* (New York, 2010).

Middleton, R. (ed.), *Amherst and the Conquest of Canada* (Stroud, 2003).

Morriss, R., *Naval Power and British Culture, 1760–1850: Public Trust and Government Ideology* (Farnham, 2004).

Parrott, D., *The Business of War: Military Enterprise and Military Revolution in Early Modern Europe* (Cambridge, 2012).

Pichichero, C., *The Military Enlightenment: War and Culture in the French Empire from Louis XIV to Napoleon* (Ithaca, NY, 2017).

Roth, J.P., *The Logistics of the Roman Army at War, 264 BC – AD 235* (Leiden, 1999).

Sánches, R. T., R. T. *Military Entrepreneurs and the Spanish Contractor State in the Eighteenth Century* (Oxford, 2016).

Santiago, M., *A Bad Peace and a Good War: Spain and the Mescalero Apache Uprising of 1795–1799* (Norman, Ok., 2018).

Shrader, C.R., *Communist Logistics in the Korean War* (1995).

Shrader, C.R., *The First Helicopter War: Logistics and Mobility in Algeria, 1954–1962* (1999).

Shrader, C.R., *U.S. Military Logistics, 1607–1991: A Research Guide* (1992).

Shrader, C.R., *A War of Logistics. Parachutes and Porters in Indochina, 1945–1954* (1999).

Stone, D.R., *The Russian Army in the Great War: The Eastern Front, 1914–1917* (Lawrence, KS., 2015).

Sutcliffe, R., *British Expeditionary Warfare and the Defeat of Napoleon, 1793–1815* (Woodbridge, 2016).

Sweetman, J., *War and Administration: The Significance of the Crimean War for the British Army* (Edinburgh, 1984).

Syrett, D., *Shipping and Military Power in the Seven Years War: The Sails of Victory* (Exeter, 2008).

Thompson, J., *The Lifeblood of War: Logistics in Armed Conflict* (1991).

Tilly, C., *Coercion, Capital, and European States, AD 990–1990* (Oxford, 1990).

Turner, J., *Feeding Victory: Innovative Military Logistics from Lake George to Khe Sanh* (Lawrence, K.S., 2020).

Wetzler, P., *War and Subsistence: The Sambre and Meuse Army in 1794* (New York, 1985).

Index